To Jane,

Thank you so much
great love of God and His Creation.
And your passion to teach.
May the Lord bless you.

Dan Richards.

ENDORSEMENTS

Mission Wild: One Man's Adventures in Saving the Planet is a new book on Christian environmental stewardship from someone who has known and experienced God's creation on a level few of us ever will. Part personal testimony, part adventure story, part theological treatise, this book offers a fascinating glimpse of vanishing ways of life on an endangered continent, while reminding us of God's redemptive plan for all of his creation.

It is especially uplifting to hear the testimonies of people whose lives have been transformed by Christ, and gain a better understanding of how God is at work in Africa.

Don Richards, a native South African, has spent his entire life in the bush of Africa, living adventures that most of us will only dream of, and he shares that with us here. But he also shares the biggest adventure of all: how Christ found him and transformed him into a "Fisher of Men" and an advocate for His Creation. With Richards unique understanding of the wildlife and people of Africa, this book is an exciting and welcome addition to the literature of creation care and missions.

— Scott C. Sabin
Executive Director, Floresta USA

★ ★ ★

Don Richards is loving and gracious — a man who pours out encouragement to those around him. In these pages, he shares of his life and faith in the world of ecology and stewardship. He has committed himself totally to the cause of Christian environmental education. This book is deeply personal and a living out of his convictions.

— Bob Bandringa
Environment and Resource School (EARS) former student

Mission Wild: One Man's Adventures in Saving the Planet

Published by Mission Enablers Publishing (MEP)
75-5775 Oleka Place
Kailua Kona, HI 96740

E-mail: info@missionenablerspublishing.com

Cover Design by Jane Ray, Fayetteville, AR.
Editorial services by Stephen Caldwell and WordBuilders, Fayetteville, AR
Page design by James Gilzow, Bella Vista, AR

ISBN 978 0 9763449 2 6

Library of Congress Catalogue No. 2008935697

First Edition

Printed and bound in the United States of America.

DON RICHARDS

MISSION WILD

ONE MAN'S ADVENTURES
IN SAVING THE PLANET

www.missionwild.com

CONTENTS

PREFACE

God is the Creator of this world and has given mankind the authority to look after it. Don, through his autobiography and adventures as a game ranger and later as a missionary in Africa and the Amazon, encourages Christians and the Church that Biblical environmental stewardship should be an integral part of discipleship and peoples' lives. The book also demonstrates to non-Christians God's involvement in His creation.

Principles of ecology are featured, described as God's secrets, insights into God's character. Adventure stories experienced by the author in the wilds demonstrate how life can be an exciting journey of discovery. Don also addresses the global warming debate.

The book includes chapters arranged chronologically and a mixture of Biblical teachings and personal adventure stories that go hand in hand with teachings. It also covers topical subjects such as global warming and climate change as well as sustainable community development and the need for environmental education.

Mission Wild: One Man's Adventures in Saving the Planet is written for Christians as a tool for understanding their environment, discipling them in Biblical Environmental Stewardship. Environmental Stewardship is not a New Age prerogative, but should be part of teaching in the church.

The book is also directed to non-Christians, to demonstrate God's involvement in His creation. It is not a scientific treatise but describes ecology in a simple way for the layman.

It is also useful for missionaries and people preparing for missions as an encouragement of how they can use their knowledge of ecology and sister disciplines in missions. It is also written for nature enthusiasts and people longing for adventure.

ACKNOWLEDGEMENTS

Special thanks are due ...

To Andy Huddleston, who encouraged me to write this book as a legacy of environmental stewardship teachings. Also, thank you, Andy, for working tirelessly to make the publication possible.

To my lovely wife, Nancy, without whom this work would not have been possible. She is forever an inspiration of encouragement to me. She patiently edited the first manuscript and has been my constant helpmate. Nancy contributed the sections about solar ovens, hot baskets, colloidal silver and tippy taps.

To so many of my close friends who have repeatedly encouraged me to write down my adventures in the wild, especially Andre Brink, who has carried the legacy to the Maasai in Kenya.

To Clive Walker, my co-author in *Walk Through the Wilderness*, who has graciously allowed me to use some of the script that I penned in that first book.

To all those who have contributed financially and otherwise to make the publication of the book possible so that more proceeds can go back into missions.

To my Heavenly Father, my Lord Jesus and the Holy Spirit, who have patiently put up with me during my wanderings and "slip-ups," as well as protecting me in the adventure of a lifetime.

INTRODUCTION

This book is about my rich experiences of God's saving grace and about His teaching me secrets of His creation, which have been termed "ecology" by science. I felt it important that people should be reminded of simple ecological truths that are really God's secrets that He put into nature from the beginning. We tend to take things for granted, and as we walk through creation in our daily lives, often we do not notice the miracles of the laws of creation all around us. There needs to be a wakeup call.

Ecology is the study of the household affairs of plants, animals and man, and how they co-exist in their natural state. In the realm of ecology, new discoveries are made continually. These are, in fact, discoveries of God's secrets that have always been there and that lead to a greater knowledge of Him. I propose a new terminology: ecotheology — the unveiling of God's secrets of how His creation works, revealing His character, corroborated by His Word.

I have tried to show how man's ignorance is impacting the creation negatively, resulting in consequences that affect man adversely. God led me to apply the knowledge of ecology in environmental stewardship through which a two-handed gospel may be brought to suffering nations. It draws people into a relationship with the Son while training communities to look after themselves. This is true discipleship. As communities are transformed, it is clear "the hand of the Lord that has done this."

That is why I have written from my experiences working in developing nations about the necessity of Christians presenting a "two-handed gospel." We must teach the Word of God and bring people to salvation, using the wonders of creation and demonstrating the urgency of stewarding the environment for the benefit of the people living in it.

The Bible tells us that the whole of creation was subjected to decay and death because of the sin of man and is awaiting the awakening of the church to do something about it. (Romans 8:18-22)

In the main, the church is not doing much to preach the message of bringing deliverance to the whole of creation nor of being an active participant in restoring it.

The Bible is full of God's love and concern for all of His creation and of how even the natural world worships Him. I have observed as I have taught so many years in environmental schools that students' eyes suddenly are opened, and they remark how, though having read the Bible for many years, they had not noticed these constant references to creation in the Word. They then became zealous to be involved in the care of creation. The church needs to be reminded of this. That is why I have included so many passages of Scripture.

In summary, *Mission Wild* is about wise stewardship of God's creation, which includes the pinnacle of His creation: mankind. My whole life has been wrapped up in conservation and ultimately creation stewardship and service, and this book is my story. I have tried to demonstrate through true experiences and ecological concepts that the Christian life is meant to be an amazing adventure.

I invite you to journey with me through the pages of this book, learning of His secrets and experiencing the adventure of doing so.

Chapter 1: **THE ADVENTURE OF LIFE**

"Life is not a problem to be solved; it is an adventure to be lived. That's the nature of it and has been since the beginning when God set the dangerous stage for this high stakes drama and called the whole world enterprise good. He rigged the wild in such a way that it only works when we embrace risk as the theme of our lives, which is to say, only when we live by faith. A man just won't be happy until he's got adventure in his work, in his love and in his spiritual life."
— from *Wild at Heart* by John Eldredge

T he river was over its banks, muddy, turbulent and in full spate. The Canadian-style canoe carrying the two of us was out of control. A fork ahead! Which way to go? Which was the main stream? Too late, we realized that we had made the wrong choice as a submerged tree stump impaled the canoe, breaking right through the bottom, spilling out our shoes, paddles, cameras and other supplies, plus Brian and me.

Roping ourselves together with one end tied to the canoe, we waded hip-deep to a small muddy island roughly 20 by 20 feet in midstream. We were in the middle of a swamp, infested with crocodiles – what to do next? Brian fired three shots in the air to discourage hippos and crocs and to try to attract attention in this Zululand wilderness.

Negotiating a number of cautious trips back to the canoe, we recovered our snakebite kit, a small methylated spirits stove and some food that we had packed in waterproof containers.

We were in the middle of nowhere! There seemed to be no way out. Crocodiles, malaria, starvation and exposure could see to that.

A punt came into sight out of nowhere, poled by a Tonga tribesman who had been attracted by the pistol shots. Loading his boat with the remaining supplies, we poled to his village some three miles journey through the swamp, passing numerous crocodiles.

The year was 1958. I was 28 years of age. Brian, a Swaziland policeman, was 27. Our plan was to be the first to canoe the Pongola

River in Zululand until it joined the mighty Rio Maputo in Portuguese East Africa (now Mozambique) to Lourenco Marques (present-day Maputo), a journey of some 240 miles. We were young, full of adventure and loved the wilds. We were explorers!

This was the start of my journey into trying to uncover God's secrets, the mysteries of His creation with its incredible interactions and complexities. I was not a believer in those days, and I would have many more moments when I would lose my way and be out of control in life's journey. But it was a beginning.

Chapter 2: BEGINNINGS

"Before I formed you in the womb I knew you."

— Jeremiah 1:5

I was born in 1930 in the district of Semarang, Java, in the Dutch East Indies, my mother, Henrietta, a Hollander and my father, Norman James, an Englishman. Due to the Great Depression of the 1930s, my parents lost almost everything they had and decided to pack up the family – Rob, the eldest, Jimmy, the middle brother, and me, the youngest – to return to England. Travel in those days was by steamship, so we traversed the Indian Ocean, the Suez Canal, the Mediterranean and hence to England. Here my father looked for work whilst my mother took me to the Swiss Alps as a place the doctors thought would possibly cure my chronic asthma. Rob and Jim were enrolled at a boarding school in Worthing, Sussex.

My one year living like Heidi in the beautiful Swiss Alps was a memorable experience, but it failed to get rid of my asthma. In 1935 my mother and I journeyed through Germany en route to England. Whilst staying over in Heidelberg, Hitler and his storm troops repossessed the Rhine region, which had been under French and British mandate after the First World War. All foreigners were ordered to leave the country immediately. This was really the forerunner of the Second World War.

Back in England we learned that my father had shipped to South Africa to follow leads of employment there. So as a family we packed again and boarded the Carnavon Castle for a two-week

> One-third of Earth's population — about 2 billion people — are potential victims of the creeping effect of desertification (*def: the process of becoming desert*), making it "the greatest environmental challenge of our times."
>
> — United Nations University Study, 2007

sea trip to Cape Town. As I stepped onto South African soil, my asthma left me, never to return! It was the first of God's miracles in my life, even though I did not know Him.

Thus began my life in Africa, a journey that was to have many ups and downs, joys and sorrows, victories and mistakes, but that I believe God had planned to bring me back to His will for my life. I wasn't going to realize this until 38 years later, but still it was a journey that taught me much of His secrets, the marvelous inter-relationships of His creation. God, the Creator of the universe, was teaching me, often the hard way, though I did not have a personal relationship with Him.

Chapter 3: **EARLY DAYS**

"I will instruct you and teach you in the way you should go. I will guide you with My eye upon you."

— Psalm 32:8

My boyhood days in South Africa were memorable, though I did not grow up in a Christian family.

We settled first in Johannesburg, where my dad had employment as sales manager for Albion Motors, a large truck dealership. Rob, my eldest brother, went to King Edward VII High School, where he excelled in sports and Bisley rifle competition shooting, becoming eventually the champion shot amongst all the Witwatersrand schools, winning many cups and medals.

Jimmy and I attended King Edward VII Primary School.

We were not well off and could not afford many luxuries, but we did have parents who loved us. My early days in Johannesburg were marred by a number of incidences.

We had not been there long when, whilst crossing the street near our home to collect a loaf of bread from the horse-drawn bakery cart, I was run over by a motor vehicle. As I lay there, the loaf of bread squashed as flat as a pancake, a large crowd gathered and traffic was held up. My mother came screaming out of the house, and I was rushed to the hospital by the person who had hit me. I was treated for shock, scratches and bruises, as well as having my right arm dressed where a tire had grazed off a large area of skin.

The second incident occurred about a year later when I was nearing the end of standard one (first grade). I came down with double pneumonia and nephritis, a kidney disease. Taken to hospital, I lapsed into critical condition. For more than a week I was unconscious and not expected to live. But apparently God had other plans. I slowly recovered and, after six weeks in the hospital, I was taken home.

THE WAR YEARS: 1939-1945

Shortly after my illness in Johannesburg, Dad was appointed as manager of the Durban branch of Albion Motors. We moved down to the coast and rented a house with a large piece of land some 11 miles inland where it was cooler.

In 1939 Great Britain declared war on Nazi Germany after Hitler had invaded Poland with his blitzkrieg. Rob joined the South African Air Force and became a pilot, flying Beaufort fighter-bombers in the North African campaign. There he was awarded the Distinguished Flying Cross twice. This medal, apart from the Victoria Cross, was the highest award an airman in the then British Empire could receive.

Jimmy attended Durban High School and then after a few years faked his age to join the South African Navy. He was seconded to the British Royal Marines as a commando and was captured by the Japanese in Malaysia, where he spent a year in prison camp before escaping and making it back to safety.

I attended the Durban Preparatory High School (D.P.H.S.) for five years. These years saw both the happiest and saddest experiences of my young life. Dad and mom bought a house in Cowies Hill, where we kept chickens, turkeys and a horse, Wapiti, which Jimmy and I had bought and fed out of our pocket money. I would ride for miles to fetch a sack of fodder for him. One day, Wapiti ran off whilst I was collecting teff grass from a farm five miles away. I had to trudge all the way home carrying a full bag over my shoulder, to find Wapiti waiting for me there! I was only 11 years old.

This same year my mother died of double pneumonia. It was a huge blow to me as we were particularly close. Just the two of us had lived together for a year in the Swiss Alps while I fought asthma, and we had always shared our thoughts with one another. When she died, it left a great sense of emptiness in my life. Another factor that contributed to moments of sadness was that I missed my brothers.

I had recently celebrated my eleventh birthday in August 1941, when Japan attacked Pearl Harbor on Sunday, Dec. 7, 1941. Japan attacked the islands of the Pacific, China, Malaysia, and Burma and was threatening Ceylon, now called Sri Lanka, and the East coast of Africa. Japanese submarines attacked shipping in South African waters, whilst the German pocket battleship, the Graf Spe, threatened the South African coastline.

South Africa, as part of the British Empire, had joined the Allies in fighting against Germany, Italy and Japan, and there was compulsory call-up to serve in the armed forces. With South Africa at war, its coastline was very vulnerable. There was compulsory blackout every night. I and a group of other young boys acted as runners for the air raid wardens. Air raid shelters were built at schools and air raid practice drills were held each week. At homes, air raid shelters and slit trenches were dug.

Dad joined the South African Coastal Defense Corps, a unit made up of men too old to join the armed forces serving overseas. His company of soldiers often held maneuvers near our home, which lay in a wild, rural area. A group of us boys between the ages of 11 and 14 had formed firm friendships, and we participated in these maneuvers as runners, taking messages from one platoon to another and camping out in the wilds with the troops.

Since the scout masters were away fighting in the war, the scouting movement in South Africa formed "Lone Scouts." Under a central commissioner of scouts, boys were able to form their own patrols, run their activities and pass tests for badges through experts in their various fields. I became leader of the Bat Patrol, and we would often attend camps with other patrols of Lone Scouts in the province of Natal.

Back at home in the afternoons, our games would be games of war, where we stalked each other with air guns, using hard green lantana seeds instead of pellets. We were young, wild and determined to fight the Japanese when they came!

In the meantime, the Suez Canal had been closed, and all Allied war shipping made the journey round the Cape of Good Hope and up the east coast of South Africa en route to the war zone in the Far East. Consequently there were always huge convoys of troop ships and equipment protected by warships anchored off Durban harbor, the largest seaport in southern Africa. These ships were mainly waiting for docking berths to collect fresh food, water and fuel. There were always thousands of Allied soldiers, sailors and airmen wandering the streets of Durban on shore leave. For us youngsters it was a great education to meet so many different nationalities at war.

Durban folk have always been known for their gift of hospitality, and before Mom died, we would often have two or three servicemen of different nationalities spend the weekend with us, enjoying a comfortable bed and home-cooked meals. We would be all ears as they regaled us with stories of foreign places.

Looking back, even going to school was exciting in those primary school years. I had to ride our horse to the nearest bus terminal three miles away. Leaving the horse there in a friend's field, I would take a bus to the city, then an electric tram to school.

I was very fond of sport and played cricket and rugby for the school's first team, as well as doing well in athletics. I performed pretty well academically, also.

BACK TO JOHANNESBURG: 1946-1948

In 1945 Dad remarried, and I attended Durban High School for a year before moving back to Johannesburg due to dad's reposting there with his firm.

My last three years of schooling were spent at King Edward VII School, where I did well academically as well as playing rugby for the school. In my last year, I was elected as a school prefect and head of a schoolhouse. I was also appointed as a second lieutenant in the school cadets and guard commander for the annual Armistice Day Parade. K.E.S. had lost the most pupils of any school in South Africa during the First and Second World Wars.

Most of my school holidays were spent with my best friend Benjie Cawood, whose parents farmed in the bushveld on the banks of the Limpopo River. These holidays in the bush further whetted my appetite for the great outdoors and all things wild.

Chapter 4: **THE INTERVENING YEARS, 1949-1970**

T he saying goes that "a rolling stone gathers no moss." During the period of my life after leaving school at the end of 1948 until I joined the Natal Parks Board in 1971, I often could be described by this analogy. I later discovered that God used this period to train me for a future I had not considered. Those years have proved to me that God had his hand in my many jobs, providing experiences I could draw upon in the culmination of His plan for me.

After leaving high school in 1948, I joined the Mercantile Marine as an apprentice deck officer. Joining ship on the City of Johannesburg, the flag ship of the Elerman and Buchnall Line, I was to spend one and a half years learning the arts of navigation at sea and loading and off-loading techniques on this 12,000 ton cargo liner that plied the seas and ports along the southern and eastern shores of Africa and around Mauritius, all in the Indian Ocean. The ship then returned to its home port, Liverpool, England, after off-loading and loading in European ports such as Rotterdam and Antwerp and circumnavigating the whole British Isles, visiting most of the ports in England, Scotland Ireland and Wales. Looking back, it was a great adventure for a teenager and expanded my worldview. However, I would get moments of intense homesickness for Africa and couldn't wait to get back again.

After my stint of sailing the ocean blue, I resigned and on returning to South Africa, I traveled to Southern Rhodesia, now Zimbabwe. There I became a learner farmer on a vast ranch amongst the granite kopjes near Rusape in the Eastern District of Rhodesia.

Apart from cattle ranching, my main occupation was growing tobacco and maize. Tobacco at the time was Southern Rhodesia's main export. It was a demanding job that entailed seed bed duties, as well as preparation of big tracts of land, stumping out trees, cutting them into cords of wood to be later used in the curing process and eventually plowing and disking the land in preparation for the planting season.

Planting season came with the first rains followed by ridging and cultivating the lands. As the tobacco leaves ripened, we would pick a few from the bottom of each plant and during the season work up to the tips as the leaves ripened. The tobacco leaves would be stacked in large crates and transported to the tying sheds by tractor and trailer. It was all hustle and bustle as full crates arrived. Tobacco was tied in sheaves to sticks and then layered in the cooking barns that consisted of tall, tiered buildings across which the tobacco sticks were stacked up to the roof.

Each barn had metal flues running through it connected to an outside oven where wood was burnt as fuel. Sacks were laid on the floor of each barn and these were drenched with water to cook the tobacco as the heat was increased. Once the leaves changed color, usually to a golden yellow, the moisture would be removed and temperatures increased gradually every two hours until 160 degrees C when the color became "fixed." After fixing, steam was introduced into the barn to give moisture to the leaves to make them pliable and not brittle. The tobacco was then layered into stacks and covered with sacks awaiting grading time.

> In a 100-year period, a water molecule spends 98 years in the ocean, 20 months as ice, about 2 weeks in lakes and rivers, and less than a week in the atmosphere.
>
> www.lenntech.com/water, 2008

This season of curing was probably the most tiring, as we had to sleep in the barn sheds, checking the temperatures every two hours day and night. We got very little sleep.

Grading season was the most colorful time. Graders stood at rows of grading tables waiting for the boys and girls to bring heaps of tobacco to each sorter. The grading had to be exact. When enough leaves were sorted into grades, they were taken to women sitting on the floor who tied the leaves into hands. From there, the sheaves were taken to a baling or pressing box. After pressing to a set size, each bale was wrapped with brown paper and sacking with the appropriate label sewn on. The bales were now ready to be taken to the auction market in Salisbury where brisk bidding could either bring a good price or disappointment.

I did not receive a monthly salary during my farming days, but received free board and lodging, some pocket money and a small percentage of sales at the end of the season.

I was very fit doing manual work alongside the African workers. I have always believed in leading by example, never expecting anyone to accomplish anything unless I could do it myself. During weekends, if I was not on farm duty, I played rugby, cricket and tennis for the district, competing in inter-district leagues. I also would spend hours during any free time exploring the huge wild area in which the ranch was situated. We discovered ancient ruins built during the same era as the famous Zimbabwe ruins, a mystery even to this day. We discovered a huge cave that extended the whole width of a granite kopje. Its entrance was guarded by a stonewall fortification, and inside we found grain storage bins and a stream of clean water. We later learned locals probably used this cave to hide during an inter-tribal war.

I also made some pocket money by hunting and shooting doves and rock hyrax that the Africans would buy to eat.

If I had known then what I know now of the evils of tobacco smoking, I would not have entered this occupation, but I did learn a great deal about farming as I progressed from learner to assistant manager to manager. I have been able to use this knowledge in the ministry that the Lord has chosen for me.

I eventually moved to another farm in the Karoi area on the bank of the mighty Zambezi River, and was there when the huge Kariba Dam was built. During this time I not only managed the tobacco side of things but also beef cattle and big acreages of maize (corn).

It was a lonely place, forty miles from the little town of Karoi.

The owner of this farm was verbally abusive, and those were not happy days. At one time I came down with a bad case of malaria whilst the owner was away. The Armstrongs, friends who farmed close to the town of Karoi, heard of my plight and took me into their home to care for me. I was delirious for three days, but they nursed me back to health over the following weeks.

BACK TO SOUTH AFRICA

After spending seven years in Southern Rhodesia, I returned to Durban in 1958 and found employment with Lever Brothers in the African marketing division.

My job was to travel with a small team of Zulu men in a cinema van that carried not only a projector lit by carbon arks and using a daylight screen, but also stocks of different soaps and washing powders.

I actually enjoyed this time of my life, as it enabled me to travel most of the rural and wild places in southern Africa, including Zululand, Swaziland, Lesotho, then known as Basutoland and Botswana. I met many interesting people from all walks of life and different races and made many friends. It was a broadening experience, teaching me about different cultures.

It was during this period that I met Brian Campbell in Swaziland and had that exciting adventure of canoeing from Gollel in Swaziland to Lourenco Marques in Portuguese East Africa.

After this experience I landed a teaching job with a private primary school in Natal. Here I taught most academic subjects, as well as physical education, and ran a wildlife club. During school holidays, I would organize camping trips for the students into various game parks in southern Africa. I made good friends with many game rangers and wardens, learning more and more about the wild and its creatures.

Throughout my childhood and into my teens and adulthood, I have loved the wilds. Often as a youngster I would stalk animals through the African bush, pretending that I was an American Indian on the hunt. I would read exciting books such as *The Last of the Mohicans, The Young Fur Traders* and other books about American frontiersmen. I learned to read signs where animals or humans had passed and loved analyzing them -- probably not all correctly!

When I was 41 years old, I was given the opportunity to join the Natal Parks Board, an organization that managed the game reserves in Natal and Zululand in South Africa. It is an esteemed body that has been responsible in saving the white rhino (square-lipped rhinoceros) from extinction. After a few months of training I was appointed to lead Wilderness Trails in the Umfolozi Game Reserve, home of the white rhino. It also contains a wide variety of wildlife including lion, leopard, elephant, black rhino, buffalo, giraffe, zebra, wildebeest, crocodile, antelope and smaller mammals. As I led trails on foot throughout this beautiful reserve and camped out amongst the wild animals, I learned about the trees and other plant life, as well as enjoying the incredible beauty of God's creation. It was the beginning of an extensive learning curve for me, enabling me to stretch my curious mind and satisfy my hunger for adventure.

THE WILDERNESS LEADERSHIP SCHOOL

After a few years on the Natal Parks Board as a game ranger and wilderness trails officer, I was invited by Ian Player, brother of Gary Player, the golfer, to join the Wilderness Leadership School as a trails officer.

Ian, a noted South African and international conservationist, was at that time deputy director of the Natal Parks Board. He had also founded the Wilderness Leadership School as a means of educating people, young and old, on the importance of wilderness and of enabling people in this changing world to experience and appreciate it.

Ian realized that unless a large body of well-informed, conservation-orientated leaders existed, irreplaceable natural resources would seriously diminish. Only by giving present and future leaders the stimulation of a wilderness experience would they come to appreciate the needs and laws of the natural environment in the face of advancing technology.

The school's trails were different in some ways from the parks board experience. The trail was a teaching course open to groups of six to eight trailists (a term used for wilderness hikers in South Africa) made up of either senior school students or groups of adults who were involved in business or government. International groups were also accommodated. Trailists would carry their own packs, help in the cooking, take turns in night watch, and have times of solitude. Both the wilderness areas of the Umfolozi Game Reserve and of Lake St. Lucia in Zululand were used as outdoor classrooms.

> God speaks to all of us in many different ways, yet it's through nature that He so easily grabs our attention.
>
> — Tri Robinson, *Saving God's Green Earth*

I had the privilege of joining an elite body made up of men such as Hugh Dent, Jim Feely and Barry Clements, the first wilderness trails officers of the Natal Parks Board who also had joined the Wilderness Leadership School.

The following chapters relate what I have learned through experience over the years about the intricate workings of God's creation. They tell about how I came to the Lord, saved by grace, to be used by Him to pioneer Youth With A Mission's Environmental and Resource Stewardship program. These Environment and Resource Stewardship Schools (EARS) train missionaries going into the developing nations.

Chapter 5: **AN INCIDENT WITH POACHERS**

Barry Clements and I were camped at the Ngabeneni Wilderness Trail Camp with a group of six schoolboys. The campsite, situated at the base of Ngabeneni (known as the Hill of Refusal) in the Umfolozi Game Reserve, has an interesting historical background.

In the early 1800s, Shaka was in the process of subduing clan after clan to form the mighty Zulu Kingdom. In his time he was considered a brilliant military strategist and has often been referred to as the Black Napoleon.

Fighting engagements between clans and tribes were often a light-hearted affair before the advent of Shaka. The women would sit on opposing hillsides, making baskets and engaging in idle chatter whilst a champion warrior from each side engaged in combat. If there were a total confrontation, the warriors on both sides would throw their spears at each other, which ensured that there were always more spears to throw.

Shaka changed all that. He did away with the throwing spear and fashioned a short stabbing spear that was nearly all blade. He also made huge shields for warriors to hide behind. Shaka was reputed to be ruthless and cruel, and he instilled a military discipline that has been unsurpassed in the annals of African history.

During his campaign to establish the Zulu nation and empire, his major foe was the Ndwandwe tribe in what is now the Umfolozi Game Reserve. Evidence of their grinding stones can often be found whilst hiking through the wilderness area.

Imagine the scene as the Ndwandwe occupied the hill flanked on three sides by steep cliffs, plunging down to the winding Umfolozi River. The fourth side sloped gently to the plains below and away from the river. As the Zulu impis (regiments) approached from across the river, the Ndwandwe protected the gentle slope, as this would seem to be the logical ascent for any attacking force. The tops of the cliff edges were lightly defended by the Nwandwe.

Shaka had other ideas. Whilst making a feint with a small force up the gentle slope, he sent his impis across the river to scale the cliffs. The Ndwandwe put up very little resistance against the fearsome sight of Zulus scaling the cliffs like a swarm of ants and fled in terror.

Shaka claimed the area and declared it his personal hunting ground to protect the elephants that were being hunted for their ivory by early white settlers who were making inroads into Zululand. Zulus named the hill Ngabaneni, which means "The Hill of Refusal" as the Zulus refused to give up and routed the Ndwandwe who initially refused to abandon it.

Barry and I took a group of schoolboys to the hilltop one morning. Here we had a magnificent view of the wilderness stretched out before us. A mixture of bushveld savannah and open, bushy country stretched away from the banks of the White Umfolozi, adorned on either side by giant sycamore fig trees, fever trees and an assortment of riverine vegetation. It was wintertime in Zululand, the dry season, and the river itself was bone dry and filled with sand.

As we sat there, spotting game through binoculars, we espied a group of some twenty poachers with their dogs hunting game on the plains below. Using simple weapons, they encircled a variety of animals, spearing and clubbing them to death.

Crouching low beneath the skyline so as not to be observed, we descended the hill, crossed the dry riverbed and walked stealthily through the vegetation in single file to where we estimated the poachers would be.

After a careful approach of about half an hour, we heard the sound of voices coming from the riverbed. Hiding in a clump of reeds, we saw the men cutting up the meat on the riverbed itself. We also noticed that all their weapons were stacked against trees on the far bank. They were quite oblivious of our presence.

Barry and I were not sure whether we should involve the boys at all. We reasoned though, that as they were all dressed in khaki, which resembled our uniforms, we would be seen as a considerable force. Further to minimize any danger to the boys, we decided we would come out all screaming at the top of our voices, firing our two rifles in the air to add to the confusion. It worked! There was immediate panic amongst the poachers who fled in all directions. We managed to capture three of them and reasoned that we could find out from them where they and their colleagues came from.

We collected all the weapons that had been left behind and as much meat as we could carry and buried the rest. Seven animals had been slaughtered.

We set off in single file towards the nearest ranger outpost some four miles away. Walking quietly along with the three poachers in our midst, we were suddenly charged by no less than three black rhino, which resulted in the whole party, boys and poachers climbing quickly and sharing available trees. After a while, the rhinos snorted off, and we continued on our way.

On reaching the outpost, we were able to radio headquarters at Mpila, which sent rangers to intercept the main body of poachers using the information given us by their captured companions. The whole group were subsequently taken to court, tried and given light sentences.

All in all it was a great adventure for the schoolboys who, I am sure, will never forget the incident. We understood, from living amongst the Zulus, that any evidence we gave in court should be geared toward leniency for the poachers, because the communities surrounding the game reserve were starving from a prolonged drought and famine.

Today there are different policies in force for protected areas, which allow the local people to benefit in some way from them. This enables them to be defenders of those protected areas rather than despoilers.

Chapter 6: **THE UMGENI VALLEY PROJECT AND ENVIRONMENTAL EDUCATION**

During the latter days of leading wilderness trails, I saw that if there were natural areas set aside for environmental education, children would grow up with a sense of the importance of conserving natural resources. They could turn over-exploitation into wise stewardship, which might spill over into different fields.

I shared my dream with many trail groups around the campfire during those Umfolozi nights. These chats did not fall on deaf ears, and in early 1974, the Wilderness Leadership School gave me the go-ahead to pioneer environmental education in Natal, South Africa.

I was given the use of a pioneer homestead in the Karkloof mountain forests, not far from a private game reserve called Umgeni Valley. We called the program "Joint Venture," a combined project of the Wilderness School, the Wildlife Society of Natal and the Natal Hunting Association.

I was joined by conservationist Roland Jones, whose major interests were botany and a love

> The growing need for agricultural land accounts for 60-80% of the world's deforestation. A change in governments' and peoples' attitudes can prevent or reverse desertification. For example dryland populations can prevent desertification by improving agricultural practices in a sustainable way.
>
> — www.greenfacts.org

for canoeing and adventure. During this time we used the Karkloof temperate forest, famous for its yellowwood trees and clear mountain streams, as well as the nearby Umgeni Valley Game Ranch, which was bushveld. Here the Umgeni River plunged over the well-known Howick Falls and continued through the valley below in a series of rapids and pools.

Joint Venture soon became the "Umgeni Valley Project" when

the Wildlife Society of Natal bought the 650-hectare property. We subsequently took charge of the game ranch and the Wildlife Society's environmental education program. Through the years this operation increased to become the largest environmental center in South Africa. Some 12,000 school children now attend the Umgeni courses annually.

Here the students learn ecological principles through field studies, equipping them both for school science studies and for becoming good stewards of natural resources. Through hiking trails combined with outdoor camping, it also became a great adventure for city children of all races. Students sleep out under the stars, experiencing the beauty of sunrises and sunsets. They experience heat, cold, scratches, getting wet and tired. Some love it, some don't, but in the end the majority come out with changed attitudes toward the world around them, realizing the need to conserve wild country and urban environments, as well.

In 1971, I had married Barbara, and together with our two children, Kim and Shaene, we experienced the Parks Board and Wilderness School days. We moved to Umgeni Valley to take charge in 1974. It was during those Umgeni days that we started attending a local Methodist church. During an evening Bible study, I gave my life to the Lord. Barbara was to do so later.

It was then that I realized how I had been influenced by New Age thinking, very prevalent in conservation circles. I knew there was a God out there, but I had been worshipping the creation and not the Creator. I had forty-five years behind me trapped in worldly thinking and had accumulated a great deal of trash in my life. Little did I realize how, as John Eldredge says in his book *Wild at Heart*, in my early Christian days, I still harbored a traitor within me.

He writes: "However strong a castle may be, if a treacherous party resides inside (ready to betray at the first opportunity possible), the castle cannot be kept safe from the enemy. Traitors occupy our own hearts, ready to side with every temptation and to surrender to them all." (John Eldredge quotes John Owen from *Sin and Tempta-*

tion in his book *Wild at Heart*)

I had been severely wounded as a child and during my teen years and, as many Christians unfortunately do, I held on to many of my old habits.

Paul gives us his famous passage on what it is like to struggle with sin.

"For we know that the Law is spiritual, but I am carnal, sold under sin. For what I am doing, I do not understand. For what I will to do, that I do not practice; but what I hate, that I do. If, then, I do what I will not to do, I agree with the Law that it is good. But now, it is no longer I who do it, but sin that dwells in me. For I know that in me (that is, in my flesh) nothing good dwells; for to will is present with me, but how to perform what is good I do not find. For the good that I will to do, I do not do; but the evil I will not to do, that I practice. Now if I do what I will not to do, it is no longer I who do it, but sin that dwells in me." (Romans 7:14-20)

Now as a Christian, I experienced this same struggle. I was still doing those things of the flesh and not of the heart. This eventually led Barbara and me to get a divorce. It was only years later when I rededicated my life to Christ and was baptized in the Holy Spirit during my DTS (Discipleship Training School) with YWAM, that I realized that my flesh was not my true self and learned it must be crucified.

Previously I had not been discipled, and the devil knew this. It took much soul-searching and good counsel from others for me to pluck up courage to bare my heart, forgive those who had wounded me and ask forgiveness of those I had wounded. But that is what I had to do and did.

ENVIRONMENTAL EDUCATION ACROSS THE CURRICULUM

During the Umgeni days, I began to think of how good it would be if a whole year of a child's life were spent studying normal school disciplines in the outdoor classroom of the environment, using the schoolroom for preparation and consolidation. A nearby Christian school heard about my ideas, and I was invited to put together a curriculum and work with a class of standard five pupils.

Environmental education is a process by which an individual develops a responsible lifestyle in sympathy with the total environment. It should include the natural and man-made parts of his world and their interdependence, as well as the ecological, economic and cultural processes that influence human lives. The general aim of education is the transfer of knowledge and the perpetuation and advancement of a culture. It should deal with the total child and should lead him to adulthood within the context of his culture, preparing him to make decisions in the face of changes. It is important for a child to gain knowledge of the environment and its basic role in sustaining life, developing concepts that will help him address the complexity of the world around him.

> There was a moment when I looked on the wilderness. Saw and heard the story of all things in it.
>
> — Don Richards, from *Reflections in Wilderness*

Environmental education should be interdisciplinary in approach, encourage active participation, examine major environmental issues in the learners' immediate surroundings and encourage individual responsibility.

Children need an education that relates to their understanding of how their environment supports them and how they affect the environment.

In education, children usually learn in a school setting with their disciplines in separate boxes. Later they must think of how to integrate those disciplines and apply them to the world. It is better, in my opinion, to take children out of the four walls of the school, and to facilitate learning by experience, using the natural world as the classroom.

The model that I put in place takes a central theme encompassing several disciplines. Beginning in the classroom, children do reference work and learn the necessary information. They are introduced to the study, divided into groups and given group assignments. They work on the assignments in the field for several days. Back in school, they consolidate through journals and teach-backs. The teach-backs include research and public speaking skills that enhance confidence. They encourage initiative and problem-solving, helping children to become environmentally literate.

Through environmental education across the curriculum, outside studies are reinforced by classroom teaching instead of classroom teaching being reinforced by outdoor learning. It uses living examples through field research and is followed up systematically. Each child's talents are developed instead of trying to make an assembly line from prototypes. The teacher acts as a facilitator and guide, learning alongside students. It includes process, research, discovery and adventure, developing self-reliance and self-discipline. Furthermore, the students enjoy school and realize how important it is to look after God's creation as they learn through a Biblical perspective.

The model that I introduced into this Christian school in South Africa is still operating successfully after twenty-six years.

Chapter 7: **NATURE AND GOD — SUSTAINABLE COMMUNITY DEVELOPMENT**

"Let a man so consider us, as servants of Christ and stewards of the mysteries of God. Moreover it is required in stewards that one be found faithful."

— 1 Corinthians 4:1-2

E nvironmental stewardship has become one of the most significant issues and concerns of our time. Institutions and communities worldwide are endeavoring both to assess and correct damage to the environment and to learn how to prevent future pollution and destructive, even catastrophic, environmental problems. As man impacts his environment, the state of the environment also has a direct effect on mankind. Christian leaders who are responsible for aspects of projects in living environments should be cognizant of today's environmental problems and practice stewardship based on a Biblical Christian worldview.

Unfortunately, I believe that, in general, the church of Jesus Christ has not recognized this crucial need and has ignored God's first command to mankind – to "tend the garden." (Genesis 2:15) It is not a New Age issue or prerogative. As Christians we need to acknowledge that God created the universe, and we need to redeem His creation and the world and everything in it. We should not ignore the world that is really there, made by a Person who is really there. Let us read in Psalm 136:3-9

Oh, give thanks to the Lord of lords!
For His mercy endures forever:
To Him who alone does great wonders,
For His mercy endures forever;
To Him who by wisdom made the heavens,
For His mercy endures forever;
To Him who laid out the earth above the waters,

For His mercy endures forever;
To Him who made great lights,
For His mercy endures forever;
The sun to rule by day,
For His mercy endures forever;
The moon and the stars to rule by night,
For His mercy endures forever.

Then again in Psalm 24:1 David wrote, *"The earth is the Lord's, and all its fullness, the world and those who dwell therein."*

If we truly love God, why do we break God's heart by desecrating His creation and ignoring man's suffering that results from this neglect?

Our function to rule and have dominion over the earth grows directly out of our position, which is that we are made in the image of God.

"Then God said, 'Let Us make man in Our image, according to Our likeness;

Let them have dominion over the fish of the sea, over the birds of the air, and over the cattle, over all the earth and over every creeping thing that creeps on the earth.' So God created man in His own image; in the image of God He created him; male and female He created them." (Genesis 1-26-27)

And then in Genesis 1:28, God continues his instructions for mankind to be fruitful and multiply, to fill the earth and subdue it; to have dominion over the fish of the sea, over the birds of the air, and over every living thing that moves on the earth.

As Christians we are supposed to be God's ambassadors, His wise stewards. There is a clear responsibility from God in Genesis 2:15, where the Bible tells us: *"Then the Lord God took the man and put him in the Garden of Eden to tend and keep it."*

Deuteronomy 10:12-14 points very strongly to the fact that everything belongs to the Lord. It gives His requirements of us and His instructions to us:

"And now, Israel, what does the Lord your God require of you, but to fear the Lord your God, to walk in all His ways and to love Him, to serve the Lord your God with all your heart, and with all your soul, and to keep the commandments of the Lord and His statutes which I command you today for your good? Indeed heaven and the highest heavens belong to the Lord your God, also the earth with all that is in it."

In view of the above, we see our duty to care for creation in both the great commandment (Matthew 22:37-40),

"You shall love the Lord your God with all your heart, with all your soul, and with all your mind."

and the great commission (Mathew 28:19-20),

"Go therefore and make disciples of all the nations, baptizing them in the name of the Father, and of the Son, and of the Holy Spirit, teaching them to observe all things that I have commanded you; and lo, I am with you always, even to the end of the age. Amen."

As Christians we are called to a high standard (doing all as unto the Lord). Our daily lives are to reflect our complete love for God and our love for people. If we carelessly participate in the degradation of God's handiwork, then we certainly are not loving God to the full. In the same way, a lack of concern for the ailing creation points to a lack of concern for the people, since a hurting creation also means a hurting people.

Introducing people to the gospel of Jesus Christ is done effectively when we demonstrate Christ's love for them in a meaningful and tangible way. For large numbers of people in our world, their biggest daily challenge is meeting basic needs of food, water, clothing and shelter. It makes sense that if we abuse the environment around us, it will not support us and God will not bless us.

It was Jesus Himself who said,

"For I was hungry and you gave Me food, I was thirsty and you gave Me drink; I was a stranger and you took Me in; I was naked and you clothed Me; I was sick and you visited me; I was in prison and you came to Me." (Mathew 25:35-36)

and He carries on to say,

"Assuredly, I say to you, inasmuch as you did it to one of the least of these my brethren, you did it to me." (Matthew 25:40)

We should be pioneers in the care of mankind and his environment. Here are some facts that impel us to environmental stewardship in the developing world:

■ More than 852 million people in the world are malnourished. (*Food for the Hungry*, 2007)

■ 6.5 million children under the age of five die every year as a result of hunger. (*Food for the Hungry*, 2007)

■ Only a small percentage of hunger deaths are caused by starvation, but are the result of chronic under-nutrition, which weakens the body's ability to ward off diseases prevalent in poverty-stricken communities. (*Time*, March 14, 2005)

■ Children's intellectual development is severely affected by lack of protein.

■ Virtually every country in the world has the potential of growing sufficient food on a sustainable basis. Fifty-four countries do not produce enough food to feed their populations, according to U.N. standards, nor can they afford to import the necessary commodities to make up the gap. Most of these countries are in sub-Saharan Africa. (FAO 2001)

■ The United Nations Development Program estimates that the basic health and nutrition needs of the world's poorest people could be met for an additional $13 billion a year. (UNDP 2001)

■ Americans spent more than $5,177,716,000 on dieting in January and February 2007. If the trend continues they will spend more than $31 billion on dieting in 2007. (Worldometer, 2007)

■ 18,000 people die of hunger every day. (*Food for the Hungry*, 2007)

■ 1,800 acres of productive land is lost every hour through soil erosion. (Worldometer, 2007)

■ 64 million metric tons of topsoil erodes from farmland every day. (Worldometer, 2007)

■ 40,350 acres of desert land was formed due to mismanagement every day in 2007. In sub-Saharan Africa, 20 percent of land has turned into desert. Worldwide deserts have increased by 150 percent in the past 100 years so that almost 50 percent of the earth's surface is desert or nearly desert. (Worldometer, 2007)

■ 1.0 percent of the land area of the earth is under cultivation.

■ A million or more children and babies die each year from disease related to inadequate water and sanitation, and hundreds of millions more suffer. (Worldometer, 2007)

■ 1.3 billion people lack access to clean water; and 2.4 billion live without decent sanitation. 28 percent of Africa's population does not have safe water sources.

■ 27,000 children die each day of preventable environment-related illness, accounting for 18 percent of diseases in the developing world.

■ 47 percent of people in Africa live on less than $1 a day.

■ Nearly 40 million people are HIV positive worldwide. WHO/ U.N. estimates that 4.3 million new HIV infections occurred worldwide this year (2007) and at about 2.9 million people died of AIDS-related illnesses.

Whilst fulfilling our prime objective of evangelism and discipleship, that is, the Great Commission, we have a responsibility concerning the first commission in Genesis 1:28 — *"fill the earth and subdue it"* — and again in Genesis 2:15 — *"Then the Lord God took the man and put him in the garden of Eden to tend and keep it"*.

Clearly, God handed to man the stewardship of managing (keeping, protecting, looking after) the resources of this planet. But man's disobedience, abuse, and violation of God's laws have polluted our lakes and rivers, expanded our deserts, poisoned the air of our cities, denuded our forests and catchment areas, and brought about poverty, disease, and huge city slums. The impending ecological disaster that the scientific community is so concerned about was predicted by the prophet Isaiah nearly 3,000 years ago.

"The earth mourns and fades away. The world languishes and fades away; The haughty people of the earth languish." (Isaiah 24:4)

And Hosea 4:1-3 reads, *"Hear the word of the Lord, you children of Israel, For the Lord brings a charge against the inhabitants of the land: There is no truth or mercy or knowledge of God in the land. By swearing and lying, killing and stealing and committing adultery, they break all restraint, with bloodshed upon bloodshed. Therefore the land will mourn; and everyone who dwells there will waste away. With the beasts of the field and the birds of the air; even the fish of the sea will be taken away."*

Here we see a direct connection between man's moral standards and the environment.

The Sentinel Group has documented a number of cases in which the environment is restored when people repent and turn back to God. An example that occurred in the village of Amolonga, Guatemala, is documented in Transformations. When people turned from idol worship and devoted themselves to prayer and fasting to God, farmers in the area began to have amazing results, with production of crops increasing by 1,000 percent. A friend of mine, Scott Sabin, the director of FLORESTA, who sent some of his staff to the area to witness this, said they testified to the incredible productivity of the land. Another well-documented occurrence, reported in the video, Let the Sea Resound, occurred in Fiji in 2000-01 when islanders turned away from demonic rituals and toward God in prayer and repentance: Fruit trees began to bear fruit again, a poisonous stream became fresh and pure, and a whole coral reef was restored.

In 2 Chronicles 7:13-14, God responds to King Solomon's prayer, saying, *"When I shut up heaven and there is no rain, or command the locusts to devour the land, or send pestilence among My people, if My people who are called by My name will humble themselves and pray and seek My face, and turn from their wicked ways, then I will hear from heaven, and will forgive their sin and heal their land."*

In Deuteronomy 28, Moses outlines the blessings and curses resulting from God's people's choices to listen to and obey Him. These include the productivity of the land and the well-being of livestock. In this age of scientific reasoning, we Christians may tend to disregard the cause and results in the physical realm of our relationship to the Creator and redeemer of the earth. God is still sovereign over the earth.

The question is: Should we as believers ignore or even denounce environmental causes? Decidedly not! To our shame, we have allowed New Age and similar humanistic causes to assume the lead in environmental action. But the responsibility has always been with God's people, so we are just as accountable today as our Biblical ancestors. The state of the environment is deeply related to man's morality and his concept of God. God has never rescinded His first commission.

An example of how people's view of God and the world affects the environment is the way some rural Africans believe they have the right to chop down trees despite the consequences (of which they are often oblivious). Peter Ollimo on our staff at YWAM Mbita Kenya, who has a degree in agriculture and who was born and raised in the area, has seen the impact of this mentality. What used to be

> In developing countries, infant mortality in drylands averages about 54 children per 1000 live births, twice as high as in non-dryland areas and 10 times the infant mortality rate in developed countries.
>
> — **www.unep.org**, 2006

forests and good grasslands and very good soil now has degenerated into semi-desert. One day he inquired of a local charcoal producer, who he remembered selling charcoal 20 years before when Peter was in secondary school. As the gentleman walked by with a donkey pulling a cart full of charcoal, Peter asked him if he was having trouble finding trees to make charcoal. "Oh, yes," the man replied, "it is hard to find trees these days. I have to go way up into the mountains to find trees." Peter asked him how many trees he had cut down in his lifetime. He replied that there were so many, he could not keep count. Peter then asked him if he had ever planted any trees. His answer was no, that was God's job to do!

We need to elevate the degree to which we value and cherish human life, and we need to elevate the degree to which we value and cherish God's creation by developing an interest and concern for the natural resource issues of a hurting creation. Remember, He was well pleased with it when He created it. This in turn will enhance our ability to identify with the struggles of indigenous people and enhance our ability to share the gospel of Jesus Christ. The love of Jesus Christ encourages us toward a compassionate response to people's real needs, putting into effect a two-handed gospel.

As Christians we need to pay close attention to problems affecting the land and creation. The world needs missionaries with expertise in a variety of natural resource fields, such as sustainable agriculture, forestry, ecology, appropriate technology, fisheries management, water resources, environmental education and other disciplines. We can recall a familiar phrase used in community development circles: "Give a man a fish, feed him for a day. Teach a man to fish, feed him for a lifetime." But that can happen only if he knows how to sustain the fish population he relies on for food.

When Jesus died on the cross and saved us from sin, He did so not only to get us into heaven but for a far more important reason. In fact, He died and rose again in order to make us into persons who could do magnificent things for others in His name, to work through us to accomplish things He wants done in this world: To eliminate hunger, to clothe the naked, to release the oppressed into freedom and to bring justice to the downtrodden. He saved us so that through us He might bring about some of the changes that are essential if He is to make this world into the kind of world He willed it to be at creation. He wants us to learn that this abundant life can be had only through loving service to others in His name — not in getting what we want, but in doing what He wants. God has promised us life lived in abundance, and when resources flow like a stream from our hands to other hands, from our hearts to other hearts, it becomes a stream that will never run dry.

Chapter 8: **TERRITORIAL BEHAVIOUR**

"Every place that the sole of your foot will tread upon I have given to you, as I said to Moses. From the wilderness and this Lebanon as far as the great river, the River Euphrates, all the land of the Hittites, and to the Great Sea towards the going down of the sun, shall be your territory."

— Joshua 1:3-4

Whilst I was working in the wilderness areas of Zululand as a game ranger and wilderness trails officer, I began to see the relationships that exist between all living and nonliving things. Walking through the wilds week after week, year after year, I was able to observe examples of this interrelatedness.

During my treks with groups through the bush, I would carry a .458 rifle, as did the Zulu game ranger who accompanied me. Having two rifles instead of one is a safety factor when dealing with big game, and especially as we had the well-being of the trailists to consider. It was a fairly common occurrence to be charged by black or white rhino, especially when there was a baby rhino present. It was expedient, therefore, to have some good-sized trees between us and the rhinos.

Rhino have notoriously poor eyesight, and cannot distinguish what you are more than six paces away. However, rhinos have very large ears and a keen sense of smell, and should we make the slightest noise or there be a sudden change of wind direction, they would be immediately aware of our presence. Hence it was a tactic of approach, whilst studying these animals, to stay downwind from them.

The white or square-lipped rhino are not usually aggressive, whilst black or prehensile-lipped rhino usually are. The white rhino will invariably try to get out of the way. When with her young, the white rhino mother will always travel behind the baby, directing it with her horn since they live in the more open tree savannah. The young are curious and will often approach any sound you make. The problem then is that you have two rhino, one of nearly three tons plus the smaller one crisscrossing down on you. The procedure then is to get your party to stand behind large trees, whilst mother and child come careering past you. I was at one time able to stand behind a tree and pat two rhinos on the rump as they came by. On other occasions we would have to fire a shot into the ground in front of them to turn them aside.

Black rhino, however, are a different kettle of fish. They seem to be quite aggressive, and climbing a tree becomes the safest tactic. They are smaller than the white and more agile.

On one occasion, whilst I was leading a group of six adults, a black rhino charged us. Fortunately the trail party was all standing by large acacia nigresens trees, which have sharp hooked thorns. This tree is the favorite habitat and browse of the black rhino.

I was standing at the base of such a tree, rifle at the ready, whilst a young lady who happened to be a model and very proud of her long blonde hair, climbed it. I quickly glanced up and noticed that her feet were too close to the ground and encouraged her to climb higher. Unfortunately her beautiful hair became entangled in the thorns and the only way to be absolutely safe was for her to move to a sort of upside down position. The rhino soon snorted off, and we had the unenviable job of cutting her golden locks to extricate her!

On several occasions during a charge, when the animal came too close, I have prayed and it has moved at a tangent. We serve an awesome God.

As an example of the web of life that God has created, we can study the territorial behavior of the white rhinoceros. Where before it was thought that food was the primary factor controlling dominance within a species, it has been proved fairly conclusively that territory is the important factor.

As in most species, the male white rhino is territorial. One bull maintains a territory of anywhere up to 250 hectares (625 acres) in

Overlapping territories

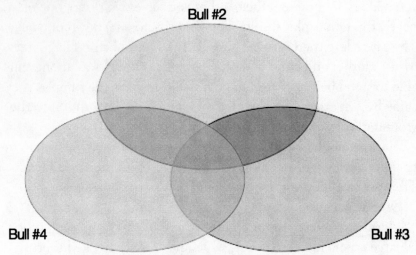

Middle shaded area = Zone of Conflict

size in the wild. There is no fixed shape to this area.

The territorial bull will mark his territory by urinating backwards in a fine spray that impregnates the scent into the ground or foliage. All rhinos urinate backwards between their legs, but usually in a steady stream rather than in a spray. In the process of marking territory, though, the bull will change to spraying. It must be understood that not all the bulls are territorial; they have to earn this dominant position. The dominant bull will also mark territory by pawing the ground and rubbing his horn and lips onto bushes or the bark of trees. Any other male of the same species then knows that that particular area is occupied. The rhino bull usually marks his territory from the center outward, like the spokes of a wheel. This often leads to confusion as it tends to overlap another bull's territory.

The territorial bull is the only male that will actually mate with the cows. Even if there are mature males within his territory, they will not mate until they have their own territory. When a territorial bull is challenged, it is usually by a mature bull, not of his own offspring, and one from outside his territory. If the old dominant bull loses, he will either stay where he is and adopt the attitude of a subordinate bull, or he will move out and become a loner. The cows are not territorial and will wander from territory to territory until held by a particular dominant bull when the females are in estrus.

It would appear that all living things revolve around this territory factor. Animals from insects to man are all territory conscious. Man builds fences around his territory to mark it or fights to protect his country.

In some birds, especially migrating species, the male calls out his territory from tree to tree in an endeavor to attract females. Birds will fight off invaders of their own and other species to defend their territory.

Dogs, whether wild or domestic, mark their territory by urinating and are usually upset when another male canine invades it.

Hippo territorial males spray their feces either at corners of their territory or along the whole boundary line, even in the water.

THE MIDDEN

A number of mammals, including antelope and rhino, consolidate their toilet by defecating on the same spot in a particular area. This collection of dung is called a midden. I have observed small and immense middens caused by the white rhino.

The territorial bull, when defecating on a midden, once again shows his dominance by scraping his dung over the entire midden, so that his smell overpowers all others. Middens with hollows in the middle usually indicate a territorial bull's presence.

White rhino will often gather in small herds; I once counted 20 of them together. The black rhino, however, seems to be more solitary or to gather in much smaller groups. It is not unusual to see evidence of black rhino dung on a white rhino's midden. The difference in the dung is very marked. White rhino are grazers, eating only grass, whilst the black rhino is a browser, feeding on leaves, twigs, branches and even thorns. Their dung is reddish in color and usually contains evidence of bark.

God's wondrous ways are evidenced in middens by the way they show interacting cycles in His creation. The fresh dung attracts insects that come for food and to breed. The dung or scarab beetle, for instance, collects balls of dung, rolls them away and buries them. It has been discovered that these balls store the beetle's eggs and are the source of food for the new grubs as they hatch out. The balls are often rolled a considerable distance and in this manner the soil becomes enriched over a wide area. Meanwhile, the dung in the midden becomes broken up, making the nutrients more easily absorbed into the soil during weathering.

Flies and other insects are also attracted to the midden, and lay their eggs in the dung. The insects and their larvae attract birds and insect-eating mammals. The birds in turn attract the smaller wildcats. Monitor lizards also come to the midden for food. One can observe a rather comprehensive food web revolving around a midden.

The book of Joshua contains much on the territory of the children of Israel as their inheritance. Likewise we as the children of God have our inheritance in the territory of the Kingdom of God.

Chapter 9: **THE WEB OF LIFE**

"I urge you in the sight of God who gives life to all things."
— 1 Timothy 6:13

"O Lord, how manifold are Your works!
"In wisdom You have made them all.
"The earth is full of Your possessions."

—Psalm 104: 24

If you take time to walk through God's creation, whether it's the African bushveld or savannah, forests, jungles, deserts, seas, lakes or rivers, you will observe that God's natural environment lives in a continuous tension. There are inter-relationships between all living and nonliving things. Nothing lives in isolation; everything is part of a community.

A tree standing alone on a hill is alone only in its distinctness from other trees. In the soil below the tree, there are many other organisms the tree relies on for growth, not to mention, air, water and sunlight.

A forest is, therefore, a complex of atmosphere, climate, rocks, soil, water, plants and animals ranging from microscopic organisms to large mammals – a complex of living things occupying a particular area is known as a "biotic community." This community, in combination with inanimate parts of the environment — soil, water, sunlight and air —forms an "ecosystem." If man harvests a forest commercially, he should consider not only the trees but also the rest of the ecosystem to which the trees belong. All the living resources are inter-related and function as a dynamically balanced whole.

The term given to the study of the environment is "ecology." The word "ecology" is derived from the Greek word "oikos," which means "house." Ecology can be thought of as the study of the household affairs of plants and animals and how they co-exist in their natural state.

What we need to be about in today's world, addressing man's disregard and degradation of the environment, is restoration ecology. This means trying to return the environments damaged by man's impact, as much as possible, to an original state in which they can be sustainable.

WHAT IS ECOLOGY?

It is the study of the household affairs of plants, animals, and man and how they coexist in their natural state.

Economics = the management of the household.

Ecology and economics should be companion disciplines.

Illustration by Don Richards

Water, soil, air, plants, animals and man depend upon each other for existence and replenishment. Invisible threads or "strands" bind living things to their environment. The complexity of these strands can be illustrated in many ways. For example, soil is rendered fertile by animal droppings and rotted vegetation. Millions of tiny organisms seas, rivers and lakes provide food sources for fish and other aquatic creatures. Shells, many thousands of years old, provide man with oil from which petroleum is made. The air we breathe is continually being renewed by the action of plants.

Each biotic community, or "biome," supports a varied cross-section of life. Each organism within a biome has its own pattern of existence. Organisms adapt to play a particular role in the biome. This role is called a "niche"!

A crab's role or niche in the biome is as a scavenger as it feeds on dead animal material. A lion's niche would be as a predator.

Often species will devise ways and means to make their niche more effective and convenient. A spider's web, for instance, is constructed as a net designed to entice and catch the spider's prey.

Young saltwater fish live close to the edges of lagoons or estuaries. Their presence provides food for countless fish, birds and reptiles, as well as man.

A hawk feeds upon mice, which rely on grass seeds for nourishment. The hawk may support a number of parasites such as lice. On the death of the hawk, it provides food for numerous organisms of decay. The mouse population also sustains other predators like owls and wild cats. The grass is eaten by other herbivores. Food chains, therefore, become interlaced into complicated food-webs.

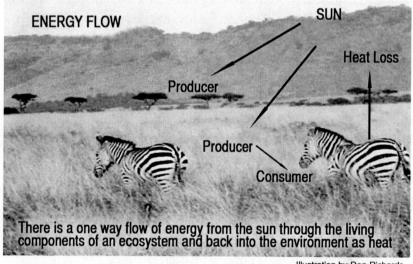

ENERGY FLOW **SUN**

Heat Loss

Producer

Producer

Consumer

There is a one way flow of energy from the sun through the living components of an ecosystem and back into the environment as heat

Illustration by Don Richards

In the food chain and the flow of energy, plants are the only organisms that have the ability, through their leaves, of capturing energy from the sun, 93 million miles (150 million kilometers) away. With the aid of chlorophyll, water and minerals that are absorbed by the roots and transmitted to the leaves manufacture carbohydrates and sugars that enable the plant to grow. This process is known as photosynthesis.

A locust draws energy from the leaves of a plant. The locust is then eaten by an insect-eating bird that obtains energy from the locust. A bird of prey captures the insect-eating bird, and the energy is transferred from one living thing to another up and down the chain.

The amount of energy available in sunlight is large in relation to the amount actually captured and used by a biotic community. Through reflection and radiation, sunlight and energy often escape capture. In each energy transfer from plants to herbivores and carnivores, some energy is lost. A lot of energy is expended just in the capture and eating processes. Therefore, the quantity of green plant material produced needs to be large in relation to the quantity of herbivores that feed upon it. The energy stored in plant carbohydrates and proteins cannot be turned into an equal amount of animal tissue. A great amount is lost in the form of heat, chemical conversions during digestion, and so on. Similarly, the quantity of herbivores needs to exceed the quantity of carnivores that depend on the plant-eaters for their energy supply. One kilogram of antelope meat cannot sustain a lion. Energy enters the system and is finally lost as heat is radiated. It is not cyclic as it has to be constantly captured to continue the system. The flow of all other life essentials is cyclic. It is never lost.

Patterns of different food chains occur in the following: In water habitats tiny water plants are eaten by fish such as tilapia, which are eaten by predator fish like catfish, which in turn are eaten by crocodiles.

In the African bushveld and savannah the grass is eaten by zebras that are preyed upon by lions. Lions, when they get old, are often killed and eaten by hyenas, which are predators as well as scavengers.

Because nature wastes nothing, energy is utilized to the very last resource. For instance, when lions have finished with their kill, hyena, jackals and vultures take the rest, the hyena even going as

TROPHIC LEVELS

Definition: A trophic level of an organism is its position in a food chain. Levels are numbered according to how far particular organisms are along the chain from the primary producers (plants) at level 1, to herbivores (level 2), to predators (level 3), to carnivores or top carnivores (level 4 or 5). Fish at higher trophic levels are typcally of higher economic value.

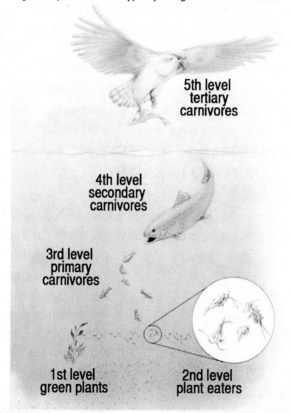

5th level
tertiary
carnivores

4th level
secondary
carnivores

3rd level
primary
carnivores

1st level
green plants

2nd level
plant eaters

far as consuming the bones.

Smaller things like flies feed off the decayed meat and lay their eggs in the carcass, so that young larvae can also derive nourishment. Borers devour the remaining bones, which eventually disintegrate into the earth, enriching the ground in the process. All the animals, from the lowest form to the large predators, enrich the soil in the form of manure. So the food chain becomes a cycle as well.

The various steps in a food chain are called trophic levels, a term that has much the same meaning as feeding levels, but is more inclusive as it incorporates the acquisition of energy by plants, a process that cannot be envisaged as "feeding." The various trophic levels in an ecosystem are as follows:

■ Producers: Normally photosynthetic plants that capture radiant energy from the sun and convert it to stored energy.

■ Primary consumers: Usually herbivorous animals feeding directly on green plants.

■ Secondary consumers: Carnivorous animals, whether predators, parasites or scavengers. All feed on primary consumers.

■ Tertiary consumers: Animals feeding on secondary consumers.

■ Higher-order consumers: All other predators, parasites and scavengers feeding chiefly on tertiary consumers.

ECOLOGICAL OR FOOD PYRAMID

As energy is passed along a food chain or web, a large proportion is lost during metabolic processes at each trophic level. When this phenomenon is represented graphically it results in a stepped pyramid shape.

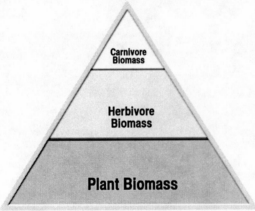

The units used in a food pyramid could be of energy expressed in calories, of biomass expressed as dry weight, or of numbers in terms of individuals. The pyramid shows that a relatively large biomass of either plant or animal material is required at low trophic levels to support a lower biomass of animals at higher trophic levels. Animals at high levels are near the apex of the pyramid (that is, low in numbers) while animals at low levels are near the base (that is, high in numbers). Any increase in numbers at a higher level will increase demands on organisms at a lower level. I emphasize this point because increased human populations in an area will certainly bring a burden on the lower trophic levels such as plants and animals. This is what has happened in the areas around Lake Victoria in east Africa.

CYCLES IN THE ECOSYSTEM

Besides the transfer of energy between trophic levels, other transfers also take place. Organisms require materials such as water, nitrogen, carbon dioxide and a number of other elements. The more or less circular paths of chemical elements passing back and forth between organisms and environment are known as biochemical cycles. The rate at which vital elements become available to biological components of the ecosystem is more important in determining primary and secondary productivity than flows of solar energy. If an essential element or compound is in short supply in

The carbon cycle

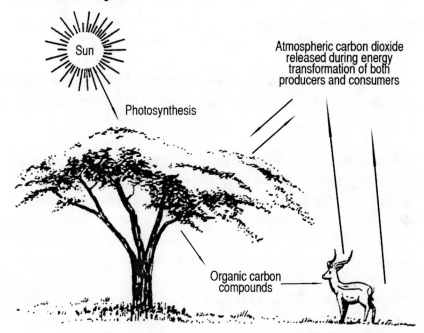

Sun

Atmospheric carbon dioxide released during energy transformation of both producers and consumers

Photosynthesis

Organic carbon compounds

Illustration by Don Richards

terms of potential growth, the substance may be said to be a "limiting factor." The productivity of an entire ecosystem can be limited by the lack of one material. Thus water limits the desert ecosystem, and nitrogen or phosphorus often limits ocean ecosystems and agro ecosystems. In ecosystems relatively undisturbed by man, the rate of cycling remains fairly constant, and any changes that do take place usually take a long time. An example of a cycle in the ecosystem is the carbon cycle.

MAN IN THE ECOSYSTEM

It is widely accepted that all natural systems are interconnected. As man is directly and indirectly dependent on the physical and biological components of the ecosystem; he not only takes out from the ecosystem but also makes "inputs." Because of the inter-connectedness of systems, man has the potential to modify and alter not only his immediate ecosystems, but eventually all. Acid rain, desertification and possibly global warming are examples of this.

Man continually makes scientific and industrial advancement, but in many cases man has not advanced at the same rate in managing his world and its resources, God's creation.

Rivers and lakes have silted up through soil abuse and incorrect farming methods. Usable clean water and good soil become scarcer each day from silt, pollution and overgrazing. It has been estimated that less than 1 percent of the earth's water supply is now of use to man.

Atomic power can be very useful, but already man has polluted the earth with radio-active dust. Unwise use of insecticides has resulted in poisoning plants, animals and human beings. Factories have spilled their effluent into rivers, lakes and seas. Forests, the earth's "lungs" have been felled indiscriminately and burned, causing increased greenhouse gases. Then there is the increased carbon emission from fossil fuel use!

What does the Word of God say? There are several warnings:

Ezekiel 34:17-19 says, *"And as for you, O My flock, thus says the Lord God: 'Behold, I shall judge between sheep and sheep, between rams* and goats. *Is it too little for you to have eaten up the good pasture, that you must tread down with your feet the residue of your pasture- and to have drunk of the clear waters, that you must foul the residue with your feet? And as for My flock, they eat what you have trampled with your feet, and they drink what you have fouled with your feet."*

Jeremiah 2:7 says, *"I brought you into a bountiful country, to eat its fruit and its goodness. But when you entered, you defiled My land and made my heritage an abomination."*

Perhaps God's most graphic warning comes in Revelation 11:18: *"The nations were angry, and Your wrath has come, And the time of the dead, that they should be judged, And that You should reward Your servants the prophets and the saints, And those who fear Your name, small and great, And* **should destroy those who destroy the earth**.*"* (emphasis added)

I believe that it is important for us not only to heed these words from Scripture, but to take close note of and understand God's secrets, the natural laws that He put in place when He created the world and the heavens. Without an understanding of these laws, we cannot have the wisdom to care for the creation.

Chapter 10: GOD'S COVENANT RELATIONSHIP WITH CREATION

The sun is the source of all light and energy on earth. Without it, living things would not exist. It powers the growth of plants and is the precursor to energy flows and food chains. It provides the light and warmth essential for all living creatures, including human beings, created in God's image.

Sunlight is key to the process of photosynthesis in plants. A plant in the dark withers and dies. Likewise, Jesus the Son of God is the source of both our physical and spiritual life. John 1:3-5 says, *"All things were made through Him, and without Him nothing was made that was made. In Him was life and the life was the light of men."*

In Psalm 84:11 we read, *"For the Lord is a sun and a shield."*

God-given light and life can reach our hearts by way of:

■ The written Word of God.

■ The Good News about Jesus Christ – His life, death, resurrection, and ascension.

■ The Holy Spirit.

■ The realm of nature, which reveals the attributes of God -- His unity, beauty, power and wisdom.

How do we respond to the light? The choice is ours. We need to repent, allowing the light to penetrate our souls. In the Christian's life, the response to the light is the same as for a plant: It's growth. In 2 Corinthians 3:18 Paul teaches, "But we all, with unveiled face, beholding as in a mirror the glory of the Lord, are being transformed into the same image from glory to glory, just as by the Spirit of the Lord."

Our character is changed and transformed into the very character of God when we are continually exposed to Him.

A tree may be flooded with light, but unless the temperature is above 40 degrees Fahrenheit (15 degrees centigrade), there will be little growth and no fruit. Low temperatures produce stunted trees. As Christians who are exposed to the light of Jesus, we, too, need warmth so that we are not stunted. In the Christian life, the counterpart of physical warmth is love or affection. 1 John 4:8 says, *"He who does not love does not know God, for God is love."* Knowing God merely in an intellectual way is not the answer. We need to know Him intimately as a loving Father, a Daddy, returning the love He first had for us. As our love grows, the Holy Spirit fills us to produce His fruit, like a healthy fruit-bearing tree.

Eighty percent of the earth's water is surface water. The other 20% is either ground water or atmospheric water vapour. Of all the water on Earth, only 2.5% is fresh water. Fresh water is either groundwater (0.5%), or readily accessible water in lakes, streams, rivers, etc. (0.01%)

www.lenntech.com/water, 2008

A tree is transformed into a light-storing, light-giving body. When a tree is felled and used as firewood, it is transformed through burning to give off heat and light. Malachi 3:2 speaks out, *"But who can endure the day of His coming and who can stand when He (Jesus) appears? For He is like a refiner's fire and like launderers' soap."* His light within us is released through the refining process.

We read in Revelation 22:1-5, *"And he showed me a pure river of water of life, clear as crystal, proceeding from the throne of God and of the Lamb. In the middle of its street, and on either side of the river, was the tree of life, which bore twelve fruits, each tree yielding its fruit every month. The leaves of the tree were for the healing of the nations. And there shall be no more curse, but the throne of God and of the Lamb shall be in it, and His servants shall serve Him. They shall see His face, and His name shall be on their foreheads. There shall be no night there: They need no lamp nor light of the sun, for the Lord God gives them light. And they shall reign for ever and ever."*

What a promise!

My life has been one of continual transformation, drawing me out of the dark into His wonderful light. It is a continuous process, but of one thing I am sure – in the Son I am rich!

Having looked at the interdependence of living and nonliving things in the physical world in the last two chapters, let us see how we can apply that concept to man and his physical and spiritual world.

Let us examine some Biblical truths regarding Christ as Creator and Redeemer.

The very beginning of the fourth Gospel, John 1:3-5 tells us plainly, *"All things were made through Him."*

In the book of Colossians, chapter 1, verses 15-20, we see again a very strong statement of truth showing Christ as Creator:

"He is the image of the invisible God, the firstborn over all creation. For by Him **all things** *were created that are in heaven and that are on earth, visible and invisible, whether thrones or dominions or principalities or powers.* **All things** *were created through Him and for Him. And He is before* **all things***, and in Him* **all things** *consist. And He is the head of the body, the church, who is the beginning, the firstborn from the dead, that in* **all things** *He may have the preeminence. For it pleased the Father that in Him all the fullness should dwell, and by Him to reconcile* **all things** *to Himself, by Him, whether things on earth or things in heaven, having made peace through the blood of His cross."* (emphasis added)

The apostle Paul emphasizes here the superiority of Christ over all creation. Christ is the one who created all things. It also indicates that Christ is the head of His own body, which is the church. The sovereign Creator of the universe, as the Head of the church, provides leadership and oversight over creation. No wonder that he is jealous over it. The question is, are we?

In the well-known passage of Scripture, John 3:16, we read, *"For God so loved the world that He gave His only begotten Son, that whoever believes in Him should not perish but have everlasting life."*

The early Greek translation of the Bible uses the word "cosmos" to denote the world. God so loved the whole of His creation that through His only begotten Son, the whole of creation would be saved and redeemed. This seems to tie in with what Paul writes in Romans 8:19-22 – *"For the earnest expectation of the creation eagerly waits for the revealing of the sons of God* (the church). *For the creation was subjected to futility, not willingly, but because of Him who subjected it in hope; because the creation itself also will be delivered from the bondage of corruption into the glorious liberty of the children of God. For we know that the whole creation groans and labors with birth pangs until now."*

In the first three chapters of Genesis, we can surely see that God was well pleased with all of His creation, including man. Adam was given the command in Genesis 2:15 to *"tend the garden and look after it."* God's call to mankind through Adam was to be obedient to Him, but Adam and Eve, tempted by the devil, disobeyed God by eating from the tree of the knowledge of good and evil, and so fell. They were banned from the garden, and the whole of creation lost man's caring stewardship.

In Romans, the apostle Paul presented the last Adam (Jesus Christ) as undoing the damaging effects of the first Adam. All that the first Adam wrecks, the last Adam restores (Romans 5:12-19). We can visualize this in the chart below:

THE TWO ADAMS	
First Adam	**Last Adam**
Disobedience	Obedience
Condemnation = Death	Justification = Life
Cursing of relationships with God, others, cosmos	Blessing of relationships with God, others, cosmos

When Adam fell, the threefold relationship unraveled. He disobeyed God (Genesis 3:1-7) and experienced distance from Him (Genesis 3:8-10). The harmony with Eve was undone (Genesis 3:11-16). The caring stewardship of the earth turned to toil in a cursed environment and estrangement (Genesis 3:17-19). Human beings became homeless, dismissed from the garden (Genesis 3:23-24).

The Kingdom of God is one of harmony and care; the opposite is a kingdom of cursing and abuse. Those who stand in the obedience of Christ have the most profound reason for practicing caring relationships and stewardship.

Now let us get back to interdependence. We know that the interdependence of believers as the body of Christ is essential. It is the same with creation; people need to fit into an interdependence with creation. To make this point, I would like to give an example of interdependence amongst a colony of termites.

In a territory (termite mound) we find four kinds of termites. These are the Queen, the workers, the soldiers and the young kings and queens. It is the Queen, about the size of your little finger that the colony revolves around. She lays all the eggs. The first eggs that hatch out will always be a large group of workers. They build the nest, gather food, and feed and tend the queen and the rest of the young as they hatch out. The next group to emerge from the eggs will always be soldiers, whose job is to defend the nest. The next batch to hatch out are young kings and queens, the flying termites. Once they are ready to leave the nest, usually after a good rain, they fly off to start new colonies. They remind me of missionaries, reaching out to plant new churches.

Now you will never see a worker trying to be a soldier or vice versa. The different kinds know where they fit in to make the colony a harmonious whole as well as to extend their kind to further fields. Another interesting phenomenon that I have observed in termite nests is that once the queen dies, the whole nest stops functioning and dies. If Christ does not rule in our lives or in our churches, we lose harmony with each other and our environment, and then we and the church become dead!

Now doesn't this remind us of Paul writing in 1 Corinthians 12 about the unity, diversity and interdependence of the body?

"For as the body is one and has many members, but all the members of that one body, being many, are one body, so also is Christ. For by one Spirit we were all baptized into one body—whether Jews or Greeks, whether slaves or free—and have all been made to drink into one Spirit. For in fact the body is not one member but many.

"If the foot should say, 'Because I am not a hand, I am not of the body,' is it therefore not of the body? And if the ear should say, 'Because I am not an eye, I am not of the body,' is it therefore not of the body? If the whole body were an eye, where would be the hearing? If the whole were hearing, where would be the smelling? But now God has set the members, each one of them, in the body just as He pleased. And if they were all one member, where would the body be?

"But now indeed there are many members, yet one body. And the eye cannot say to the hand, 'I have no need of you'; nor again the head to the feet, 'I have no need of you.' No, much rather, those members of the body which seem to be weaker are necessary. And those members of the body which we think to be less honorable, on these we bestow greater honor; and our un-presentable parts have greater modesty, but our presentable parts have no need. But God composed the body, having given greater honor to that part which lacks it, that there should be no schism in the body, but that the members should have the same care for one another. And if one member suffers, all the members suffer with it; or if one member is honored, all the members rejoice with it."

— 1 Corinthians 12:12-30

Paul goes on to say that God has appointed those in the church, some as apostles, some as prophets, some as teachers, then the different gifts of healing, miracles, helps, administrations, varieties of tongues. Each operates for the good of all.

The Hebrew word "shalom" describes all that God intends for creation. Shalom is all about wholeness: It is the peaceful beauty of natural settings where one can feel the harmony of creation working together as God planned. It includes abundance, diversity and the wise and fair use of resources. Within the concept of "shalom"

love would describe man's interaction with his fellows and his surroundings. In the same way, truth would lead to right moral decisions, and all of creation would be rejoicing before the Lord instead
of languishing. "Shalom" would also include freedom from corruption and the well-being and productivity of earth's systems: Croplands producing, trees bearing fruit, clean water, seas teeming with
life. In fact, "shalom" means wholeness. Isaiah 11:6-10 describes
God's intent. This wholeness occurs not only within the individual
person because of his relationship with God when he is in covenant; it is also tied to man's right relationship with all creation.

If we are not in harmony with the natural world, we are not
a whole. This is a process we
experience, a realization of how
we rely on the natural world and
how it relies on us. This is the
interdependence I have been
talking about. We need to be in
a process of realizing wholeness.
The theological word is "sanctification." Sanctification is the
ongoing process whereby the

> About one-half of the
> mature tropical forests —
> between 750 to 800 million
> hectares of the original
> 1.5 to 1.6 billion hectares
> — have fallen.
>
> — http://wikipedia.org

Holy Spirit works in believers, making their lives holy, separated
from their old ways to be more like God — being set apart with
and for God.

The guiding vision for our wholeness is the Kingdom of God.
Our task is bringing about God's vision for the sanctification and
redemption of all of creation. This is the creative task that God has
given us.

The Old Covenant that God gave Moses on Mount Sinai, the
Mosaic covenant, was that if you keep all the laws you will be
righteous. Many of the laws given in Deuteronomy and other Old
Testament Scriptures dealt with natural laws of how to look after
the environment. (Leviticus 26:2-5; Exodus 23:10-12)

The New Covenant that came through Christ was that He

would give us His righteousness. This covenant also includes creation.

In Jeremiah 31:31-34 we read, *"Behold the days are coming, says the Lord, when I will make a new covenant with the house of Israel and with the house of Judah—not according to the covenant that I made with their fathers in the day that I took them by the hand to lead them out of the land of Egypt, My covenant which they broke, though I was a husband to them, says the Lord. But this is the covenant that I will make with the house of Israel after those days, says the Lord: I will put My law in their minds, and write it on their hearts; and I will be their God, and they will be My people. No more shall every man teach his neighbor, and every man his brother, saying, 'Know the Lord,' for they all shall know Me, from the least of them to the greatest of them, says the Lord. For I will forgive their iniquity, and their sin I will remember no more."*

Jeremiah carries on with reference to creation in chapter 33, verse 12. *"Thus says the Lord of hosts: 'In this place which is desolate, without man and without beasts, and in all its cities, there shall again be a dwelling place of shepherds causing their flocks to lie down.'"*

God's Word carries the assurance of including creation in this new covenant in Ezekiel 36:33-36 – *"Thus says the Lord God: 'On the day that I cleanse you from all your iniquities, I will also enable you to dwell in the cities, and the ruins shall be rebuilt. The desolate land shall be tilled instead of lying desolate in the sight of all who pass by. So they will say, "This land that was desolate has become like the garden of Eden; and the wasted, desolate, and ruined cities are now fortified and inhabited." Then the nations which are left all around you shall know that I, the Lord, have rebuilt the ruined places and planted what was desolate. I, the Lord, have spoken it, and I will do it.'"*

What an emphatic statement!

Before the creation of man and woman, God declared that what had been created was good. Therefore the natural world has value to God, even apart from its relationship to human beings. Nature's worth is therefore not dependent upon what it can do for us.

We can see this clearly stated in Scripture. In Job, chapters

38-41, God exults over how He designed different aspects of the natural world and some of its creatures.

He asks Job, *"Where were you when I laid the foundations of the earth? Tell Me, if you have understanding. Who determined its measurements? Surely you know! Or who stretched the line upon it? To what were its foundations fastened? Or who laid its cornerstone, when the morning stars sang together, And all the sons of God shouted for joy?"* (Job 38:4-7)

God has implanted in animals awareness of time and seasons. He wonders that man refuses to be sensitive to His promptings. In Jeremiah 8:7, it says, *"Even the stork in the heavens knows her appointed times; And the turtledove, the swift, and the swallow observe the time of their coming. But My people do not know the judgment of the Lord."*

It is truly remarkable how birds in their annual migrations know exactly when and where to go. Swallows and swifts will come back each year to nest in the same place down to the very same house or bridge.

Then in the book of Job it is written in chapter 12, verses 7-10, *"But now ask the beasts, and they will teach you; And the birds of the air, and they will tell you; Or speak to the earth, and it will teach you; and the fish of the sea will explain to you. Who among all these does not know that the hand of the Lord has done this, in whose hand is the life of every living thing, and the breath of all mankind?"*

It seems that creation knows but that man doesn't, and because of that man causes his own suffering, and that of his environment. We see this clearly in Isaiah 24:4-6 – *"The earth mourns and fades away, the world languishes and fades away; The haughty people of the earth languish. The earth is also defiled under its inhabitants, because they have transgressed the laws, changed the ordinance, broken the everlasting covenant. Therefore the curse has devoured the earth, and those that dwell in it are desolate. Therefore the inhabitants of the land are burned, and few men are left."*

When we look at the earth today and what man has done to cause so much waste and pollution and the lack of moral standards that results in the epidemic of AIDS, we need to take heed of what God is saying to us through His word, which was the same yesterday, today and tomorrow.

We have a choice to make between life and death. We can see this stated clearly in Deuteronomy 30:19-20 – *"I call heavens and earth as witnesses today against you, that I have set before you life and death, blessing and cursing; therefore choose life, that both you and your descendants may live; that you may love the Lord your God, that you may obey His voice, and that you may cling to Him, for He is your life and the length of your days; and that you may dwell in the land which the Lord swore to your fathers, to Abraham, Isaac, and Jacob to give them."*

In the environmental context, the blessings listed in Deuteronomy 28:2-14 include blessings on the Israelites' children, the produce of their land, and the increase of their herds and flocks. God promises to give rain to the land in its season. It is of our choosing whether we obey God's commandments and keep His covenant or disregard them and bear the consequences in the environment.

God has given us dominion over His creation. To have dominion means a call to stewardship of God's possessions, for God is the true owner of creation. Idolatry in any form and disregarding God's commands breaks covenant with God, resulting in oppression and injustice, bringing destruction to creation.

To be in covenant and to have proper dominion is to understand what it means to be created in God's image. It is not an excuse for us to become little gods ourselves but rather to become more like God — to love and care for creation as He does. This obviously includes man's relationship to man.

We have been charged by God to care and look after creation. We are called to conserve and to care for and cherish all life, so that wholeness may return to creation — this is Shalom.

In Genesis 6:11-22, we see the Creator of the universe, who owes man nothing, taking one man, Noah, into His confidence. In response, Noah was obedient and faithful. God instructs him why and how to make the ark and also to take a pair or more of every living animal and every sort of food onboard. Everything that was to go into the ark was valued. None of what God created was worthless.

In Genesis 8:20-22, we read about the Noahic covenant, God's covenant with creation: "*Then Noah built an altar to the Lord, and took of every clean animal and of every clean bird, and offered burnt offerings on the altar. And the Lord smelled a soothing aroma. Then the Lord said in His heart, I will never again curse the ground for man's sake, although the imagination of man's heart is evil from his youth; nor will I again destroy every living thing as I have done. While the earth remains, seedtime and harvest, cold and heat, winter and summer, and day and night shall not cease.*"

In Genesis 9:8-17, we see that God established His covenant with Noah and all creation by making the rainbow a symbol of God's promise never to flood the earth again and as a constant reminder of His oath.

THE REVELATION OF CREATION

Thus all of creation is in covenant with God. Again consider Romans 8:16-22, where we read that all of creation is waiting in expectancy for the sons and daughters of God, the Church. All creation is impatient to see the revelation of the sons of God. It awaits man's awakening to his responsibility as God's wise stewards that it might be delivered out of bondage. That means that nature (creation) is a slave to decay and death because of man's sins.

We see, too, throughout Scripture that everything in creation praises God.

"*And every creature which is in heaven and on the earth and under the earth and   such as are in the sea, and all that are in them, I heard saying: 'Blessing and honor and glory and power be to Him who sits on the throne, and to the Lamb, forever and ever.'*" (Revelation 5:13)

In Revelation 4:11, that divine rule has its basis in creation and redemption. Let us read from Revelation 4:9, "*Whenever the living creatures give glory and honor and thanks to Him who sits on the throne,* who lives forever and ever, the twenty-four elders fall down before Him who sits on the throne and worship Him who lives forever and ever, and cast their crowns before the throne, saying: 'You are worthy, O Lord, to receive glory and honor and power; for You created all things, and by Your will they exist and were created."

The elders cast their crowns before the throne, symbolizing the willing surrender of their authority in light of the worthiness of God as Creator.

Because no one but God can create, He alone should be worshipped and recognized as sovereign.

One of the strongest statements of God's covenant with creation and its implications for justice is found in Hosea:

"In that day I will make a covenant for them with the beasts of the field, with the birds of the air, and with the creeping things of the ground. Bow and sword of battle I will shatter from the earth, to make them lie down safely. I will betroth you to me forever, Yes, I will betroth you to Me in righteousness and justice, in loving-kindness and mercy; I will betroth you to Me in faithfulness, and you shall know the Lord.

"It shall come to pass in that day that I will answer," says the Lord; "I will answer the heavens, and they shall answer the earth. The earth will answer with grain, with new wine, and with oil; they shall answer Jezreel. Then I will sow her for Myself in the earth, and I will have mercy on her who had not obtained mercy; Then I will say to those who were not My people, 'You are My people!' And they shall say, 'You are my God!'"

— Hosea 2:18-23

> These creatures minister to our needs every day: without them we could not live; and through them the human race greatly offends the Creator. We fail every day to appreciate so great a blessing by not praising as we should the Creator and Dispenser of all these things.
>
> — Francis of Assisi, 1181-1226

In closure, we should understand that mankind is different from the rest of created things because we are made in the image of God. One thing that sets us apart is our ability to reflect — we make decisions that result in either death or life, war or peace, pain or pleasure, brokenness or wholeness. Because of this ability, we can destroy the natural world or be wise stewards of it.

When Jesus died on the cross, the creation mourned; the sky darkened, the earth quaked, and the veil of the temple was torn from top to bottom. He bore the curse that came upon creation because of man's disobedience when He bore the crown of thorns on His head. Jesus our High Priest, the Lion of Judah, established the New Covenant through His blood.

The biblical call to wholeness is referred to throughout the Bible and paid for with the blood of Jesus. That is why it is so important that you and I have a right relationship with the Creator and all that He made.

Chapter 11: **A CLOSE CALL**

A pride of twelve lions were on the beach of the White Umfolozi River. Some were standing in the shallows; others lay
basking in the sun, just as house cats like to do. From our vantage
point behind a screen of low bushes and reeds, we watched fasci-
nated for over an hour, marveling at their grace, size and power.
We were privileged to be so close, about forty yards away, and they
were quite unaware of our presence.

Since it was almost sundown, I whispered to my trail party, all
businessmen, that we must be on our way if we were to get back to
camp before dark. To reduce the distance and time, I decided to
take a short cut. Unfortunately, this meant following a rhino path
that meandered through thick scrub, with very few trees. It would
be risky as the terrain was perfect black rhino habitat.

Carrying my .458 rifle at the ready, I lead the way warily through
the scrub. Then it happened! Out of nowhere a bulky projectile of
muscle and horn came charging straight at me through the bushes.
I shouted a warning to the party behind me. "Trees," I yelled. But,
alas, there were no trees to climb, and the best they could do was to
lie down individually behind bushes.

The black rhino, snorting vigorously, was only twenty paces
away and coming straight for me like an express train! I shot a
round into the ground in front of the animal, a strategy that normally turned them aside, but it just kept coming. I raised the rifle
to my shoulder, took aim to fire into its head, knowing that the
impact of a .458 bullet would drop it in its tracks. It was so close
that I could see the mucous seeping out of the corners of its eyes. It
is remarkable what detail one can remember under stress!

Finger on the trigger, I began to squeeze it as a last resort, when
suddenly it made a right-angled turn on the spot and crashed
through the bushes and out of sight.

Ashen-faced, the men got up from behind the bushes and
walked over to me, my heart pounding. They measured the dis-

tance from the turn marks to where I was standing—three paces!
Too close for comfort!

Excitedly, we discussed the incident, and looking back I now
understand that God had intervened. The Lord, whose creation we
walked through, was in control and had saved me for a mission that
would become clear to me when I gave my life to Him.

Likewise, the so-called laws of nature, really God's natural laws,
are His secrets that few people understand. People tend to walk
through the world with blinders on, oblivious to the interesting
patterns that animal and plant species follow together with the
nonliving resources in an eco-system. Some animal and plant
species compete fiercely for food or living space, whilst others
manage to work beneficially together.

> The power of God is present in all places, even in the tiniest leaf — God is currently and personally present in the wilderness, in the garden, and in the field.
>
> — Martin Luther, 1483-1546

Animal and plant species exist
through various forms of association or interaction. In the indif-
ferent state of "neutralism," plant and animal species are unaffected by
contact with the other.

Competition results in populations adversely affecting each other
in the struggle for food, nutrients, living space or other common
needs. In the antelope world, for example, nyala and bushbuck derive
their nutrition from the same browse, but the nyala, being taller, con-
sume plants that the bushbuck cannot reach. Among plants, many
compete with one another for light, space, food and water.

As people, we tend to compete with each other, as well, which is
not always a bad thing. Paul reminds us how we should be competing against ourselves to spiritually win the prize.

*"Do you know that those that run in a race all run, but one re-
ceives the prize? Run in such a way that you may obtain it. And everyone who competes for the prize is temperate in all things. Now they*

do it to obtain a perishable crown, but we for an imperishable crown."
(1 Corinthians 9:24-25)

Mutualism is the system in which the growth and survival of populations is inter-related and one cannot survive without the other. The terms "co-operation" and "symbiosis" are also employed to define this system. Many flowers rely on insects for pollination. In return, insects acquire nectar from the flowers as a main food source. A bee, for example, would never survive if it couldn't carry nectar and pollen back to the hive, the center of bee populations. The bee then travels from flower to flower leaving behind some of the pollen on the pistle, which then travels to the ovary of the plant, enabling it to reproduce. The bee and the flower are vital to each other for continued existence.

In the plant community, lichen is a combination of an alga and a fungus. The alga, which contains chlorophyll, produces food and energy; the fungus provides a "skeleton" for the alga to grow on. Both the alga and the fungus are so closely associated that botanists refer to the lichen as an individual species. In nature, neither type of algae and fungi could exist alone. Lichen is a pioneer plant, strong and capable of overcoming hard conditions to open up areas for softer communities. God also wants us to be pioneers, strong and able to overcome hardships through His power in us, preparing the way for new Christians. That is what the Great Commission is about. (Mathew 28: 19-20)

Commensalism relates to one species benefiting while the other, in "friendly association," remains unaffected. For instance, barnacles live in permanent attachment to a whale's hide but cause no inconvenience or harm to the whale.

Parasitism occurs where one population adversely affects another through direct attack but at the same time is dependent on its host. Ticks, tsetse flies and mosquitoes are common parasites, while birds harbor many parasitic insects in their feathers. Cuckoos are parasitic in nature as they do not build their own nests but lay their eggs in the nests of other birds such as robins'. The robin has no idea that it is hatching and raising a bird of a different kind. Since the young cuckoo grows faster and larger than the robin chick, it monopolizes the food, and the robin chick perishes.

In the plant world, blight is a microscopic parasite found on a host plant, which it eventually destroys. A common parasitic plant, the loranthus or mistletoe, is found clinging to acacia trees that cover vast areas of the African veldt. In the United States it attacks various species of pines, including the ponderosa pine. Another parasitic plant, striga, invades whole crops of grain in Africa. All these parasitic plants have reduced root systems modified into special absorbing organs, because they do not need to absorb water or mineral salts from the soil. The have smaller quantities of chlorophyll and either scale-like, colorless leaves or no leaves at all, since they rely on the host plant to capture the sun's energy.

This reminds me of Malachi 3:8-10 – *"Will a man rob God? Yet you are robbing Me! But you say, 'How have we robbed you? In tithes and offerings. You are cursed with a curse, for you are robbing Me, the whole nation of you! Bring the whole tithe into the storehouse, so that there may be food in My house, and test Me now in this," says the Lord of hosts, "if I will not open for you windows of heaven and pour out for you a blessing until it flows."*

If one ambles through a dense canopy-covered forest, he sees many dangling plants that look like parasites because of the way they entwine themselves in the trees. In point of fact these plants, known as "epiphytes," are not parasitic but only use other plants as mechanical supports. They do not draw nourishment from these "prop" plants. Like climbers, epiphytes abound in tropical and subtropical forests and fall into sun-loving and shade-loving groups, depending on the niche where they exist. Orchids, ferns and lichen are well-known epiphytes.

Common inhabitants of tropical and sub-tropical forests are the "stranglers," especially the species of wild figs. At first, the seeds of strangling figs germinate like epiphytes, but their roots grow down, reaching the soil. As the fig grows, more roots develop, increasing in girth and surrounding the trunk of the "'host," which is gradually killed. The fig then becomes self-supporting, and when the host tree rots away, a hollow-trunked fig tree appears.

To understand the process of how plants draw nourishment, one must look at the process that moves water and minerals from the air into root systems. Osmosis is the term used to describe the movement of water containing dissolved elements through a membrane dividing a stronger solution from a weaker one. This movement flows from the weaker solution until both solutions reach the same level of concentration. This occurs in the roots of a plant.

Soil water surrounding the particles of soil is actually a solution of mineral salts weaker than the solution in the cells of the root hairs so that water passes through the cell wall into the root hairs. The absorbed water travels from the root hairs, also as a result of osmosis, into the root's wood-vessels. Through the roots, stems and leaf stalks, it eventually reaches the leaves and then evaporates into the air.

The process of osmosis affects certain plant adaptations. For instance, the plant Scaevola thunbergii or dune disc leaf not only stabilizes dunes on the beach but is adapted to withstand salt spray from the waves — a very strong solution which would draw water out of the leaves if they were not protected by the thickness of the leaf cuticle, or external leaf layer.

The nivea plant is another one efficiently adapted to withstand salt spray because of a special silky coating on the uppermost leaf surfaces. If the coating weren't there, osmosis would occur in reverse, as the salty solution on the outside would be stronger than the solution inside the leaf. The plant would become dehydrated through excess loss of water.

Another interesting fact in the transportation of water and minerals from the air is apparent in the legume family. Certain microscopic bacteria living in small nodules on the roots of plants of the bean family. Leguminosa, including the acacia species, have the ability to extract nitrogen from air in the soil. On their own, plants cannot achieve this. The nitrogen, converted into nitrogen compounds, reacts with the substances manufactured in the green parts of the plant to form proteins — important foods for animals and man. Agriculturists use leguminous crops to increase nitrogen content in the soil.

We serve an awesome and creative God, and we often do not realize how creative. If we study the laws of nature just described, which so many scientists accept as natural laws without any intelligent design or designer, we can surely see His handiwork. It is not just a matter of chance.

Chapter 12: **ADAPTATION**

"All the rivers run into the sea, yet the sea is not full; to the place from which the rivers come, there they return again."

— Ecclesiastes 1:7

An estuary is formed where one or more rivers converge to flow into the sea, often forming lakes or lagoons. Lake St. Lucia in Kwa Zulu Natal on the east coast of South Africa, is southern Africa's largest estuary. Some 75 kilometers long and 25 kilometers at its widest point, it has been set aside as one of South Africa's pristine wilderness areas.

It is a beautiful place, a refuge for ocean fish, hippos, crocodiles and numerous birds, including the greater flamingo and white pelican. Its eastern shores are interspersed with marshes, grasslands, subtropical forest and smaller lakes. Dividing it from the Indian Ocean is a band of sand dunes mostly covered in dune forest, culminating on the beach with dunes 600 feet tall, some of the highest in the world.

An estuary is an important ecosystem, as it is made up of many different habitats and containing a wide variety of species. They are very fragile areas.

Fresh water flowing down the rivers mixes with seawater brought in with tides, and salinity varies from 0 to 35 parts per thousand from the head of the estuary to the mouth. Salinity changes according to the season. For instance, the rainy season reduces the salinity, and dry, windy seasons evaporate the water more quickly, causing higher salinity.

Depletion of the fresh-water supply to Lake St. Lucia in the late 1960s caused such extremes in salinity that it was feared the whole system would collapse. The problem occurred when some 56 rivers and streams that ran into the estuary were either diverted or dammed up by communities upstream. This, plus a particularly dry year, caused the salinity level at the head of the lake to rise three times that of sea water. The lake virtually died. For the first time in living memory, crocodiles died from dehydration, fish died, and the food chain collapsed, causing even water birds to leave in the thousands.

The fish that inhabit ocean shorelines do so according to the availability of food. Thus fish usually inhabit the continental shelf fairly close to the shore, and their movement is parallel to the coast in search of food rather than out into deep water and back.

Most fish spawn at the mouth of estuaries, and it is from here that young fish swim up into estuaries in search of food and protection. The edges of natural estuaries are covered in rich feeding grounds of zostera (eel grass) and ripia. These shallow feeding grounds are the basis of the food chain in the estuary. It is here that most life abounds, from the smallest microscopic animals and plants to crustaceans, fish, birds. Where these occur, crocodiles also abound. Islands within estuaries are important for the large feeding grounds around their fringes. When fish reach maturity, they swim out to sea again and the whole cycle is repeated.

> If thy heart were right, then every creature would be a mirror of life, and a book of holy doctrine. There is no creature so small and abject, but it reflects the goodness of God.
>
> — Thomas A Kempis, 1379-1471

Hippos in Africa are important to river and estuary life. They are nature's natural dredgers, and in rivers that tend to silt up, they move sand or silt toward the banks, increasing the current.

The loss of hippos from most of Africa's rivers has caused them to silt up or to form vast sandbars at the river mouths, creating lagoons that are difficult for young fish to enter unless they are flooded.

Hippos also create paths from swamps into rivers and estuaries, causing underground water to flow into the estuaries. They have the habit of defecating in the water, spreading the dung with their tails. The dung, which is made up of pure grass, is fed upon by some fish species. Thus hippos are vital to estuarine systems. It is sad that most African estuaries and rivers are devoid of hippos and, in fact, of most life.

Pollution is another cause of disturbance in estuaries. Some of the rivers, lakes and estuaries in the world are so badly polluted from factory effluent and human waste that they have died. Some have become so polluted that life out to sea from the river mouth has also been wiped out.

My work as a wilderness trails officer would often take me to the beautiful wilderness of the St. Lucia estuary. Here I would lead groups of trailists either walking through the mixed terrains or canoeing in Canadian canoes close to the shore. During these canoe trails we would experience the thrill of seeing hippos lying on the beach or frolicking in the water, but we always kept a circumspect distance. Now and then, skeins of pink flamingos would thread above us and groups of white pelican would sail majestically, wings outstretched, gliding inches above the water.

Crocodiles are prolific here, and we would always have to keep a wary lookout for them.

One crocodile became famous in that area. A huge beast about 18 feet long, Barnacle Bill, as he was called, would often lay on the sand or among the reeds near our campsite. His name originated from the fact that he was so old that barnacles grew attached to his body.

Barnacle Bill had no fear of man, so we were always careful to keep a safe distance from his usual resting place. On one excursion, when I was leading a group of high school students in three canoes, we espied Bill lying on the shore, enjoying the warmth of the sun. We made a wide detour toward a flock of flamingos wading in the shallows some distance away. Passing old Bill's usual territory on our way back to camp a couple of hours later, we saw no sign of him. Suddenly, out of nowhere, Barnacle Bill surfaced alongside my canoe, making us paddle furiously out of harm's way. A close call indeed!

On another occasion, whilst I was canoeing with a group, a croc bit a hole in the back of the canoe just behind where I was paddling in the stern. Fortunately the bite was above the water line and missed my rear end by inches!

The good Lord has given all living things the ability to adapt to their environment in order for them to survive successfully. For instance, a polar bear with its thick layer of fatty tissue and beautiful white fur can exist successfully in Arctic temperatures. It also blends in wonderfully with its white background of snow and ice, enabling it to stalk its prey.

Crocodiles likewise are a masterpiece of God's design, whether we like them or not. Adapted to live both in water and on land, this reptile is a master of water predation.

Crocodiles use their tails as the primary means of propulsion in the water. Whilst swimming either on the surface or underwater, their front legs are tucked in under the body, whilst the back legs are stretched out like stabilizer bars. Their streamlined bodies enable them to move through the water at considerable speed to catch catfish or mammals, as well as animals that come to the waters edge to drink.

The crocodile's eyes are on the top of its head, enabling it to hide almost unnoticed near the water's surface, with only its eyes protruding. Underwater the eyes are covered with a nictitating membrane, which allows it to keep its eyes open whilst hunting prey. The crocodile's throat is equipped with a hinged flap that closes when it opens its mouth to seize prey, preventing drowning.

There are many highly specialized forms of adaptation in the animal kingdom. When looking for lions in the African savannah, it is truly amazing how inconspicuous these magnificent animals can be — tawny-colored creatures in a maze of tawny-colored grass.

Perhaps the most dramatic example of animal camouflage is the chameleon, which changes the color of its skin continuously as it moves on different colored surfaces.

God has made the antelopes of the bushveld to move with speed when danger threatens. Predators such as the cheetah are created to run at phenomenal speed when hunting in the open savannah.

Ant bears have long, sharp claws to dig our their burrows and into termite nests to look for food. They are also equipped with long, sticky tongues to probe narrow termite tunnels. Their burrows are frequently taken over by warthogs, hyenas and porcupines, which clean the dens and use them for shelter and protection. In the breeding season, the dens afford ample shelter for the animals' litters.

Some keen-sighted animals living in tall grasslands can jump very high, enabling them to see predators at a considerable distance. The little springhare of Africa protects itself in this way. Kangaroos and wallabies in Australia jump for a different reason. Their Creator has made their front legs for plucking and feeding, and their strong rear legs for travel by jumping.

Although flying is mainly confined to birds and insects, there are a few tree-climbing animals and lizards with flaps from neck to shoulder that enable them to glide from tree to tree. I have often seen flying squirrels, in this case the southern flying squirrel from our cabin in the Smoky Mountains of Tennessee. Another example is the flying dragon, a species of lizard.

The bat is the only mammal that God has made for a life in the air. Many animals have the ability to seek food or shelter from predators in trees. Primates such as baboons and various species of monkeys are good examples of this.

Whilst conducting trails in the Umfolozi Wilderness area, I had many experiences with baboons. One day whilst another ranger and I were conducting a patrol along the riverine forest, we suddenly found ourselves in the midst of a troop of baboons feeding in a grassy clearing surrounded by huge sycamore fig trees. We approached quietly and soon were looking down on three youngsters so intent on probing the ground for grubs that they were oblivious to our presence. When they became aware of us, all three tried frantically to dig themselves into the ground to get away. They were just a jumble of bodies, legs and tails, an amusing sight indeed. As we laughed, they ran to the nearest fig tree and tried to clamber up the slippery bark, but kept sliding down in their state of panic. Eventually they reached the safety of some higher branches and were reprimanded by their mothers in the form of loud scolding and vigorous slapping.

These primates have a sense of discipline regarding their young and this was a good example of teaching them to be more alert and not to endanger themselves and the rest of the troop.

When we see the intricacies of God's creation and His design, it is impossible for me to even imagine the theory of evolution. I give you here a monkey's viewpoint!

Three monkeys sat in a coconut tree,
Discussing things as they are said to be.
Said one to the others, "Now listen, you two,
There's a certain rumor that can't be true,
That man descended from our noble race.
The very idea! It's a dire disgrace!
No monkey ever deserted his wife,
Starved her baby, and ruined her life.
And you have never known a mother monk
To leave the babies with others to bunk,
Or pass them on from one to another
`Til they scarcely know who is their mother.
And another thing! You'll never see
A monk build a fence `round a coconut tree.
Starvation would force you to steal from me.
Here's another thing a monk won't do:
Go out at night and get on a stew;
Or use a gun, or club or knife
To take some other monkey's life.
Yes, man descended, the ornery cuss,
But brother, he didn't descend from us.

— Author unknown

Years later at our YWAM base on Lake Victoria, we were plagued by monkeys eating our vegetable gardens and called in the Kenya Wildlife Services to shoot some of them. Only two were shot, and within a short time the rest of the troop came back and took the two dead monkeys away!

In addition to fish, animals such as dolphins, porpoises, walruses and seals are adapted to a life in the water. Their bodies resemble those of fish, with torpedo shapes and flipper limbs. Their nostrils and ears, when present, close when submerged. These aquatic carnivores are generally larger than terrestrial ones, ranging from 200poounds to several tons. This large size and streamlined body are believed to equip them for the cold environments they live in, as both reduce the relative amount of surface area through which body heat is lost. Layers of fat beneath the skin help them in this regard, as well as in storing food reserves. In the water they are most graceful swimmers. What a creative God we serve Who equips all life to fit into the environments He created for them.

> The day should begin and end with a sense of complete freedom that nothing belongs to you, you owe nothing and nothing is owed to you.
>
> — Don Richards from *Reflections in Wilderness*

Some birds, such as cormorants, dive and swim underwater; turtles also are adapted to life in the sea. As mentioned before, crocodiles use their tails as the primary means of propulsion in the water. Most mammals are able to swim.

The forelimbs of mammals were created to adapt to their different ways of life. For example, God made man's hands to hold and manipulate even the finest things; the forelimbs of the ape have been made to grasp to enable them to feed, climb and swing from tree to tree. Flesh eaters, that is the carnivores, have developed claws on their paws to enable them to catch their prey and to rip and tear. It has always amazed me to observe how ant bears have the ability to tear away and dig through the compact surface of termite mounds that have the consistency of concrete. As already mentioned, seals and like species have limbs that are flippers, enabling them to swim so elegantly. The camel's two-toed foot splays out to be able to traverse soft sand, whilst the more solid hoof of the horse family is adapted for running on hard ground

Birds, too, were created to have feet according to their way of life. For example:

■ Swimming feet — in the case of ducks, geese and pelicans.
■ Feet for walking and scratching — fowls and partridge.
■ Perching feet — crows, larks and most small birds.
■ Climbing feet — woodpeckers, parrots, cuckoos and mouse birds.

The beaks or bills of birds have been made to cope with their diet. For instance:

Finches have a thick bill for cracking hard seeds.

Swifts and swallows have a wide gape with hairy feathers at either end to resemble a net to enable them to catch insects in mid-air, for "eating on the wing."

Crows have bills suited for many purposes.

Eagles and other birds of prey, the kings of the air, have hooked bills for killing their prey.

Ducks have flattened bills equipped with a softening apparatus to filter food.

Curlews have long, slender bills for foraging underground.

Water birds such as the waders have different lengths of legs according to the depth of water that their food inhabits. Thus some waders feed close to the shore, whilst others wade deeper.

All biological systems are designed to maintain the so-called balance of nature. The balance is in constant flux because of the many environmental factors continuously influencing the system.

The natural process continues with biological factors that control population growth such as disease, predation, accidents, climate, food supply and stress. Where these factors are not fully operative because of man's influence, such as in fenced-in game reserves, man can give nature a helping hand by culling and relocating game through scientific methods. It is easy to imagine how the total environment would suffer if there were no biological control of the locust population, for instance.

The diversity (variety of species) within a community reflects in part the diversity of the physical environment. The greater the environmental variation, the more numerous the different species will be. This is because there are more microhabitats available and more niches to fill. This phenomenon is easy to illustrate by comparing the number of species occurring in tundra as opposed to those found in a tropical rain forest. It is vital that we conserve species diversity in an entire ecosystem and not just random parts in a particular region. A classic example was the predator control conducted in the game reserves of Zululand. In the mid-1930s, when this mistake was recognized, the practice was stopped, but not before the lion, cheetah, brown hyena, African wildcat and the Cape hunting dog disappeared from Zululand.

Mammals, as their Latin name tells us, comprise a class of animals the females of which produce milk to feed their young. Probably the class with the most general appeal, this immensely varied group of species is endowed with outstanding adjuncts to survival — in many ways superior to other vertebrates.

Being "warm-blooded," mammals have hide or skin, fur or hair, as well as fat, which insulates their circulations against changing climate, temperature and local environment. When hot, they can sweat; when cold, they can shiver. They can cool themselves by panting (like canines), or just maintain sufficient body temperature for bare existence when hibernating. Their adaptation is extremely advanced.

Mammals generally are equipped with a combination of well-developed senses: Sight, taste, smell, hearing and touch. They can also co-operate with each other for mutual survival. Being adapted as a species to underground conditions, moles are virtually blind, but their highly sensitive sense of smell guides them to food and warns them of danger. Mammals living in arid or desert conditions can go for months without water. Life in the thick bush and long grasses of the savannah calls for superb hearing as we can observe in the king-size ears of most antelope and deer. It is remarkable how the herbivores' ears are able to move independently of each other, sweeping in different directions to pick up the slightest sound. This compensates for the fact that herbivores' eyesight is not as keen as carnivores. Carnivores' ears are fixed and lack this ability, since they are the hunters and are invariably moving forward whilst they stalk their prey.

Though by no means exclusive to the class, camouflage is also a feature of the mammals' survival pattern. It is a never-ending source of wonderment to the wilderness visitor that creatures of the magnitude of elephant and giraffe can merge so expertly into the locale. Only when they move does the

> If all new sources of contamination could be eliminated, in 10 years, 98% of all available groundwater would then be free of pollution.
>
> — www.lenntech.com/water, 2008

camouflage fail to protect them. Most mammals live in a dull, gray world. Only a few — the apes, for example — have color vision. Others — usually animals that are active at night, like the cats — are famous because their eyes glow in the dark. The eyes, however, do not generate any light themselves; they merely catch up and reflect whatever light is present in the gloom.

Behind this remarkable power lies an interesting fact. The inner wall of the eye is coated with a substance called "guanine." This has a metallic luster of silver or gold and brightens dimly lit images on the retina of the animal's eye so that it can see them more sharply. I have had great fun during night walks and drives with students. As you sweep a powerful spotlight around in the dark you are able to pick out the lit-up eyes in the dark. Different species reflect a different color; for instance, crocodile eyes show up as red.

Nerve fibers end in tiny raised points on the skin in some animals and are known as touch spots. A touch spot usually has hair on it. The hairs themselves are not sensitive but act as levers to press the touch spot. Long hairs or whiskers on animals are connected to touch spots.

Some animals have bright coloring to merge into their respective backgrounds, but animals of the open grasslands are noted for their lack of markings. The under parts of these animals are usually white, or lighter than the color of the back. The brightest light comes from above and lightens the back, but throws the under parts into shadow, so that both upper and lower parts look alike. The animal's outline is thus obscured in certain lights unless it moves. It is also believed that the white underbelly reflects heat generated from the ground in open country.

If you are wandering through the wilds, it should be possible to identify mammals by their skulls when found. For instance, an animal with a long nose will have a keener sense of smell than those with short ones. The longer noses usually belong to herbivores whose keener sense of smell alerts them to the presence of predators. Eyes placed in the front of the head are of animals that have good eyesight and are usually predators. Those with eyes placed on the side of the skull are usually prey species that depend on this type of eyesight to guard against attacks from the rear.

From teeth we can see whether an animal is a carnivore, herbivore, omnivore or insectivore and whether it is a grazer or a browser. Vegetarian mammals (herbivores) are usually large and therefore need a large food supply. To achieve this large intake, the animal's life pattern is almost one long meal. The problem of obtaining food is usually more complicated for the carnivores. They have to work for their food by hunting, and they need greater intelligence to survive.

The battle for survival makes it necessary for all mammals to have offensive and defensive adaptations. For instance, antelope males and in some instances females, have horns to defend themselves.

The males also use these horns in territorial battles. Armadillos have armor plating, whilst porcupines defend themselves with sharp quills. It is not true that porcupine shoot their quills. When an enemy attacks, the quills are erected and as the attacker contacts the porcupine, the quills pierce the attacker and are removed from the owner. These quill points are like miniature fish hooks and are very difficult to remove, often resulting in the attacker being infected and sometimes eventually dying.

Chapter 13: **FRESHWATER ENVIRONMENTS**

The Black Umfolozi River was in full flood, a flash flood caused by heavy rains miles away in the catchment area, where heavy deforestation and overgrazing by cattle and goats had taken place. The river was a raging, swollen force, chocolate in color.

Paul Phelan and his wilderness trail party, returning to our base camp at Masimba, were cut off. The approaches to the causeway/bridge across the Black Umfolozi were washed away. I was standing on the side of free access to the outside world and my home at Masimba. There was no way a vehicle could cross this raging torrent. I had received a message over the radio that this party was cut off at the completion of their wilderness trail. They needed assistance, and to make it worse, one of the party, a business tycoon, was demanding to get out to meet a business commitment. The situation looked impossible.

I and some helpers tied two pieces of rope to an inflated tractor tube. They held one end on the bank while I negotiated the flood to the center of the causeway. The rest was covered by shallow water, so I was able to wade with the tube to another deep gap near the far bank. Here the current was extremely deep and strong, but we were able to shout to each other over the roar of the raging river. I would try to throw the rope to those standing on the bank and effect a rescue with the tube. Try as I might, even with a weight tied on the end, I could not get the rope over to the stranded party.

Brian Thring, another ranger with the returning party, decided to walk upriver and swim the 100 yards to the Masimba side. Brian was a powerful swimmer and managed to get across, landing on the other bank just short of the causeway. I was then able to bring him back to the center of the causeway where I was standing by using the tractor tube.

Together we finally succeeded in throwing the rope to the far bank. We yelled to Paul to tie the rope to the Land Rover and then secured our end to one of the concrete bollards on the causeway. We decided to get Brian back to the party and then ferry the businessman back to the causeway on the tube. Brian got into the tube and entered the water, but his added weight caused tube to be sucked into one of the water ducts under the causeway. I shouted to Paul to reverse the Land Rover, and out came the tube but no Brian! I felt terrible. There was no way, I thought, that anyone could survive under those conditions. I glanced downriver and, to my relief, saw Brian surface a hundred yards away and swim safely to shore. It was then that we decided to cancel the whole operation. Meeting a deadline was not worth the loss of someone's life!

This incident shows how dangerous it is to be inconsiderate of catchment areas. Our human disregard of God's original plan for clean streams and rivers through good ground cover has resulted in disastrous floods worldwide. The flooding in Bangladesh in the 1980s, where thousands of people lost their homes, land and lives, is

> Each year an additional 20 million hectares of agricultural land either becomes too degraded for crop production or becomes lost to urban sprawl.
>
> — www.unep.org, 2006

an example we should note. It is vital for us to conserve and steward our fresh-water sources and environments.

Water is essential to virtually every life form on earth and makes up about 70 percent of the human body. We are "born of water," referring to the aquatic environment of the womb, and we perish from lack of water sooner than from lack of food. In addition to the physical need for water, it has profound symbolism for our spiritual lives. God's Spirit was "hovering over the waters" during creation. (Genesis: 1:2) Water frequently has been used for ritual cleansing and is the medium for baptism.

People have taken clean water for granted, but fortunately many now recognize the need to protect and restore this resource. One author calls water "the fluid of the 21st century," suggesting that it will displace oil in its importance both to daily life and to geopolitics in the near future. (See "Everyone Lives Downstream" by William Deutsch, Ph.D. and Bryan Duncan, Ph.D. in Down to Earth Christianity.)

In many parts of the world, where drought is a common condition, the presence of rivers, streams, lakes and pans is like an oasis in the middle of a desert. In fact, many of these droughts are caused by man's mismanagement of watersheds and catchment areas.

Trekking along a dusty wilderness trail, I was constantly aware of the absence of water, and always felt a sense of relief in sighting a riverbed, a swamp or a hidden stream.

Apart from the aesthetic value of rivers and lakes and the simple need to drink water, there are millions of life forms in freshwater habitats that constantly change with the effects of the seasons, rain and sun and erosion. These weather changes and natural processes may result in water habitats turning into terrestrial habitats.

Freshwater habitats provide fascinating opportunities to study large communities of animals in the zones of lakes, ponds, swamps, rivers and streams. One zone might support many rooted plants with floating leaves like the water lilies, while another zone carries vegetation with rooted plants living in a constantly submerged state.

Freshwater environments are the homes of animals ranging from snails, dragonfly nymphs, water beetles and bugs to frogs, turtles, water snakes and the many species of fish.

Freshwater habitats may be divided into two categories: 1. Standing water, also known as levis or calm habitats. Examples of this are lakes, ponds, and swamps. 2. Running water, also known as lotic or washed habitats. This category includes springs, streams and rivers.

Freshwater habitats are prone to change from erosion, resulting in springs disappearing and ponds and lakes being filled with sediment. Streams and rivers tend to cut down to base level and change as a result of water action. When base level is reached, the current is reduced, sedimentation takes place, and a shallow, meandering river results. This is sometimes known as an "old river." This process can be speeded up considerably by man's abuse of soil, such as ploughing downhill, ploughing right to the river's edge and overgrazing in catchment areas, a situation that is all too common worldwide. The amount of silt in any freshwater system restricts the penetration of light, which in turn affects the process of photosynthesis.

In lakes and ponds, three zones are evident:

■ Littoral zone: The shallow water region with light penetration to the bottom, occupied by rooted plants.

■ Limnetic zone: The open water zone to the depth of effective light penetration. The community in this zone is composed only of plankton and other forms of microscopic life.

■ Profundal zone: The bottom and deep water area beyond light penetration, where there are very few life forms.

In streams and rivers, two zones are evident:

■ Rapids zone: Shallow water where the current is great enough to keep the bottom clear of silt, producing a firm bottom. This zone is occupied by life forms that have adapted to cling.

■ Pool zone: Deeper water where the velocity of the current is reduced, resulting in a soft, silted bottom, favorable for burrowing forms of life.

The littoral zone in lakes and ponds, because of its importance, deserves further attention:

Within the littoral zone there are three subzones:

■ Emergent vegetation: Rooted plants with leaves above the water. The emergent plants, together with those of the moist shore, provide an important link between land and water. They provide food and shelter for amphibious animals and easy entry and exit for aquatic insects like dragonflies and mayflies that divide their time between water and land.

■ Rooted plants with floating leaves. The undersurface of these lily pads provides resting places for animals and a place for attaching eggs.

A pond ecosystem

Comparison of a trophic structure of a simple terrestrial ecosystem with an aquatic ecosystem adjoining it. Basic units of the ecosystem are (1) Abiotic substances; (2) Producers; (3) Macro-consumers (a = direct herbivores; b = detritus eaters or saprovores; c = carnivores); and (4) Decomposers (bacteria and fungi of decay).

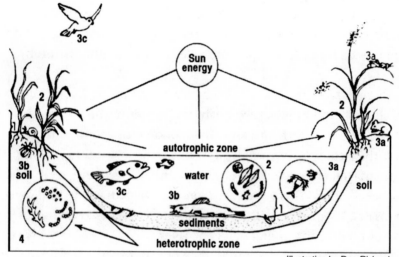

Illustration by Don Richards

■ Submergent vegetation: Rooted or fixed plants completely submerged. The leaves here are thin.

The littoral zone is home to a greater variety of animals than any other. Animals such as snails, dragonfly nymphs, flatworms, hydra and midge larvae rest on or attach themselves to stems and leaves. Snails feed on plants, but most of the other forms mentioned are carnivorous. Water beetles and bugs abound in different forms.

Amphibious frogs, salamanders, turtles and water snakes are almost exclusively members of the littoral zone. Lake and pond fish move freely between the littoral and limnetic zones, but most species spend most of their time in the littoral zone. Many species actually establish territories and breed there. The non-rooted producers of the littoral zone are comprised mostly of species of algae.

Temporary rivers and ponds that are dry for part of the year are especially interesting and support a unique community which either must be able to survive in a dormant stage during dry periods or able to move in and out of these areas, as can amphibians and adult aquatic insects.

The eggs of some of these adaptable animals are able to survive in the mud or dry soil for many months, and development and reproduction occur in a short time while water is present. This may even be seen in temporary pools on tops of mountain ranges. Some reptiles, such as the terrapin, are able to go into hibernation by burrowing into the mud.

Not all rivers are alike: Some are swift flowing and clear, others sluggish and brown. Obviously, the nearer to the source, the cleaner the river. Streams and rivers among the mountains and near their catchments usually flow over a rocky bottom. Where mismanagement and erosion have occurred, rivers that were rocky bottomed and V-shaped, now become U-shaped, wider, shallower rivers due to immense siltation. This leads to quick evaporation and ultimately the drying up of the river.

We see because of the importance of water to all life forms, and especially mankind, how vital it is for us to conserve and look after all our water resources. The Bible tells us, "As water disappears from the sea, And a river becomes parched and dries up, So man lies down and does not rise." (Job 14:11)

And then in Isaiah 41:17, "The poor and needy seek water, but there is none, their tongues fail for thirst."

These scriptures are revealing of what is happening especially to the poor in developing countries. The Lord of hosts wants us to care and help such communities by teaching them how to look after the natural resources that God has given them.

Chapter 14: **DESERTS AND THEIR PECULIARITIES**

"The wilderness and the wasteland shall be glad for them, and the desert shall rejoice and blossom as the rose."
— Isaiah 35:1

In my work in conservation and environmental stewardship, the Lord has taken me across the deserts of South Africa, Botswana, Israel and the United States of America. I have found them extremely interesting, supporting varied and very specialized forms of plants and animals.

More than a fifth of the world's land surface, an area the size of Africa, is occupied by desert, semi-desert and arid regions. Sometimes a desert is a rocky plateau, sometimes a flat, yellow plain. Here shallow lakes appear after a rainfall and then evaporate quickly, leaving a layer of glistening white salt. Other places comprise vast areas of shifting sand dunes, and elsewhere there are steep rocky slopes. In the heart of the great desert regions, rainfall is scarce and the landscape has given way to the effects of wind erosion.

The scorching heat of the day and the contrasting cold of night make deserts the most extreme environments for animal life. At dawn and dusk of each day, a desert comes alive. Mice, foxes, snakes and lizards emerge from their hideouts to seek food. Most of the desert animals have been created with water-conservation techniques. A limited number of animals are able to survive in extreme desert conditions. Of these, many live in rock crevices and under loose stones. Running, jumping and burrowing vertebrates play an important role in the desert. Lizards are common among the running forms, and mice are typical of the desert jumping animals. Rodents avoid high temperatures by remaining in their burrows until the coolness of the night. Ground temperatures during the day may reach 70 degrees Centigrade (120 degrees Fahrenheit) or more. At night, if radiation to the sky is unimpeded, the surface temperature may fall well below that of the night air. The extreme temperature variations experienced at the surface diminish with depth.

Some species of desert rodents live without water and appear to exist entirely on seeds and dry plant material. Often this plant material is stored in the burrow, where with increased humidity, the stored food takes up water and supplies the rodent's needs. I cannot ascribe this to chance, but only to an omniscient God who creates all things for their purposes and enables animals to adapt to changing conditions.

> We can gather that all the creatures of the world lead the mind of the contemplative and wise man to the eternal God. For these creatures are shadows, echoes and pictures and vestiges proposed to us and signs divinely given so that we can see God.
>
> — St. Bonaventure, 1221-1274

Lizards and snakes living in desert conditions have been created for burrowing and are able to disappear quickly into the sand. The side-winding adder (Bitis caudalis) moves horizontally and disappears rapidly and silently into the sand. It has the ability to escape predators and heat. Reptiles devise other means of protection, since they are incapable of ranging over great distances. Apart from burrowing away from heat, lizards have the ability to reflect excessive heat. It has been observed that when the body temperature of a lizard rises to an intolerable level, its skin turns lighter.

Color is useful as a means of camouflage, too. Most desert animals tend to resemble the pale colors of their environment. In the desert, every bush and rock is an oasis of animal life.

Among the inhabitants of the sandy desert floor, some animals are adapted to prevent sinking into the sand, enabling them to run or move more easily over the loose, fine sand. Some lizards, for example, have fringes on their toes. The sand grouse has feathery, tarsal toes; some rodents have well-developed hairs on the soles of their feet. The side-winding habit of the snake is an adaptation to move more easily through soft sand.

In general, only animals that tolerate dry air are best adapted to open desert country.

Reptiles and insects that can tolerate high temperatures and little moisture flourish there.

Birds survive in dry areas because their powers of flight bridge the gap between water supplies, and they also use water sparingly.

Mammals are represented by species that are relatively independent of water. Adaptations include rodents' cutaneous glands and antelopes' concentrated urine and dry feces. Many rodents such as porcupine, some antelope and anteaters can survive for months without drinking water. The moisture taken in with their plant and animal food, together with water produced in the process of metabolism, is enough to satisfy their needs.

Chapter 15: **A NEW DIRECTION**

*"And as He walked by the Sea of Galilee, He saw Simon and
Andrew his brother casting a net into the sea; for they were fisherman.
Then Jesus said to them, 'Follow Me, and I will make you become fish-
ers of men.'"*

— Mark 1:16-17

The year was 1989. My friend Roland Jones and I had successfully launched L.E.A.P. – the Lebowa Environmental Awareness Program in the North Eastern Transvaal in South Africa.
Working together with the Lebowa Conservation Department, we
were given a small game reserve to run environmental education
camps for the Lebowa school children and teachers.

The Holy Spirit was working in my life as I submitted once more
to the Lord Jesus from a backslidden position. Very clearly, through
His voice, Scripture and a series of events, He laid the tribes of the
Amazon on my heart, along with a compelling challenge to lay
everything down and to go into full-time missions.

Through another set of amazing divine appointments, at the
age of 59, I found myself as a student in a Youth With A Mission
(YWAM) Discipleship Training School (DTS) at Worcester in
the Cape Province of South Africa. At the start this was a challenge, because the rest of the students were in their teens and early
twenties, as were my leaders. It was a humbling experience, since I
had been director of a number of conservation organizations. This
was beneficial to me, and the DTS was a real blessing as it turned
me inside out, upside down and back to front! During this five-
month school, I was baptized again. The Lord opened more doors
for me to join YWAM in the Brazilian Amazon rain forest. Here I
launched A.C.A.P.—the Amazon Conservation Awareness Program as a means of reaching tribal groups and the Hiberinos River
people with the Gospel and a Christian Environmental Stewardship program.

The Lord also enabled John Kuhne and me to pioneer the
first YWAM school of Environment and Resource Stewardship
(EARS), a secondary school of YWAM's University of the Nations.
The goal of the school is to train missionaries to go into developing
countries to teach people about creation care, the care of our God-
given resources that sustain all life.

The EARS School offers YWAM staff and students the opportunity to learn how to better steward what God gives us. The
six-month course – three months lecture phase and three months
outreach – looks to equip missionaries in this field, especially those
who will minister in the developing world. There is an urgent need
for missionaries to help people access and develop resources in a
sustainable way. In fact, environmental knowledge and care is a
matter of survival for most developing countries.

Whole areas can be transformed from poverty to prosperity. The
process results in advancing the Kingdom of God and enabling
people to be able to look after themselves and to make a commitment to Him who created all.

People need also to be transformed through a biblical Christian
worldview. As God's representatives on earth we need to obey His
command to "tend the garden." Paul said, *"For we are His work-
manship, created in Christ Jesus for good works, which God prepared
beforehand that we should walk in them."* (Ephesians 2:10) By tending the garden, we not only conserve it, but also make it more
productive in a sustainable way.

Darrow Miller, in his book *Discipling the Nations*, says, "The gospel is God's total response to man's total need. We are to bring the
life and wisdom of God to bear in all of life, not just in a privatized
religious sphere. ... Our goal must be nothing short of transforma-
tional development, which impacts both man's spirit and body."

The EARS objectives aim to equip students with a clear understanding of scriptural truths regarding responsible stewardship.
Through case studies, lectures and practical activities, students gain
an understanding of current issues regarding the environment and

resources worldwide. Students emerge equipped to evaluate the status of resources in their outreach area and to engage community involvement towards restoration and sustainable development. Teaching from a biblical perspective, the ministry is a valuable discipleship tool, following the principle of a two-handed gospel (i.e. combining the good news with practical helps).

The topics of study include:
■ Stewardship in Cultural Context
■ Worldviews in Conflict; Humanism vs. Creationism; Biblical Worldview
■ The Interaction between Environment and Economy
■ Toxicology: Global Warming: Pollution: God's Solutions
■ Practical Ecological Foundations — God's Secrets
■ Community Participatory Training in Hygiene and Water
■ Agro-forestry; Soil Conservation; Sustainable Agriculture
■ Conservation; Biodiversity and Production; Micro-enterprise
■ Appropriate Technology
■ Ethnobotony; Preserving Cultural and Biological Diversity
■ Water and Health
■ Environmental Education
■ Community Transformation and the Kingdom of God
■ Communities in Crisis — TB, AIDS, Malaria and Tropical Diseases
■ Survival Skills for Tough Places/ Survival Camp
■ Cross-cultural Challenges

This YWAM school was birthed in the Brazilian Amazon in 1994. Subsequently EARS schools have been run over the years at YWAM's base on Lake Victoria Kenya, the All Nations base near Trinidad in Colorado and the Heredia base in Costa Rica. EARS internships, an eight-week missionary experience for university students interested in working in environmental issues during the summer break, are run under YWAM Madison, Wisconsin.

Chapter 16: **ON THE FRONTLINE**

"Finally my brethren, be strong in the Lord and in the power of His might. Put on the whole armor of God, that you may be able to stand against the wiles of the devil."

— Ephesians 6: 10-11

I lay in my hammock, unable to sleep. It was my first night in an Amazon Indian village deep in the jungle, three hundred miles from the nearest speck of civilization. That day I had been flown in a four-seater Cessna piloted by a Wycliffe missionary pilot to join a YWAM missionary couple in a Jarawara village.

I had mixed emotions and thoughts as I flew over the sea of green tropical forest that looked like broccoli from the air. Huge rivers were interspersed through the jungle. If we had engine failure, where would we land but in the trees! A fearful thought. Where was my faith? I was struck also by the beauty and majesty of God's creation.

We landed eventually on the tiny landing strip at the Jarawara village, a feat in itself amongst those huge trees! There I was met by David and his wife, plus men and children of the village.

As I lay sweating on that first tropical night, I heard the wailing of the local medicine man, calling in the spirits. David and his wife invited me to join in spiritual warfare all night. They explained that this calling in the spirits had not happened for a long time and that the reason for our warfare was that Satan was not pleased at my presence.

The next afternoon I joined David and the men in a football match on the landing strip. Even at my age I was very fit, able to walk many miles in a day, but the very first kick I took at that football resulted in a wrenched tendon in my knee. Time for more warfare! Within a week, though, I was able to limp into the forest to learn from that same medicine man about the trees and other Amazon plants and their uses by the local people. No longer possessed by an evil spirit, the medicine man was a totally different person from that first night. I was soon able to walk with a guide to another village fifteen miles away where I spent a few nights before walking back, learning as I went along.

As we go into different cultures throughout the world in our missionary calling, it is important to realize that spiritual forces of darkness control places and cultures, producing circumstances quite opposed to God's Kingdom and God's intentions for people. Satan is busy in the world, influencing through his lies and deception, leading people to believe that poverty, disease, corruption, misuse of the environment and apathy are their heritage.

Spiritual mapping is an important tool for those of us who are called to disciple the nations. Spiritual mapping is similar to mapping an area ecologically. As you research an area examining its history and culture and looking at current events, allowing the Holy Spirit to speak to you, core issues and focuses for prayer come to light. Knowing target areas to focus on in prayer allows for effective intercession. Spiritual mapping and survey, as well as prayer, are essential in understanding why things are as they are in communities and nations. Therefore we are given these tools of spiritual warfare and intercession by God to bring down strongholds and subsequently bring victory through Christ Himself.

> If present practices continue, by 2030 there will be only 10% of tropical forests. remaining with another 10% in a degraded condition.. 80% will have been lost and with them the irreversible loss of hundreds of thousands of species.
>
> — http://wikipedia.org

God's ways are not our ways! In missions, we sometimes believe that it is only through missionaries or the church itself that people come to the Lord. Often, though, people come to give their lives to Jesus through the Holy Spirit Himself.

We were on our way to the Kairatiana tribe, one that had

contact with civilization for at least 70 years. As far as we knew, there were no Christians there. I was traveling with Ronaldo, the YWAM base leader, and a few other YWAMers. Our goal was to investigate the possibility of creating a buffer zone between the local settlers, who were taking over Kariatiana land, and the tribe itself. The tribe needed the land to hunt and gather. We felt that a buffer zone would protect both parties' interests.

Arriving at the Karatiana village that night, we were amazed when we were invited to attend a church service that was going on. Entering the pastor's humble home that also served as the church, we were surrounded by about 35 men, women and children all praising the Lord in their own language. The pastor then shared his testimony with us in Portuguese. We were the first outsiders to hear his story.

"One year ago," he said, "I had a vision in which I was taken up to heaven and met a man called Jesus, someone I had never heard of before. I saw many people there in this beautiful place, but very few people of the forest. I asked Jesus why this was so. He replied that it was because my people worshipped demons. 'I want you to return to your village and teach your people about Me,' He said. Then He asked me what my name was, and I gave him my Portuguese name.

"He then told me my Kairatiana name, which I very well knew, but was ashamed to use. He said, 'I made you a Kariatiana, and your people a Kariatiana nation, and I desire that you teach your people about Me in your own language.' So in the last year I have been singing and teaching about Jesus in my village and now have this group around me who all believe in Him as the true God. I cannot read or write and do not have a Bible, but He tells me what to say."

This is one of the examples of the mysterious ways God is impacting indigenous people throughout nations, even without missionaries, but purely through His Holy Spirit. It also shows how God values each particular nation and culture in His world.

Chapter 17: **TROPICAL FORESTS**

"If you walk in My statutes and keep my commandments, and perform them, then I will give you rain in its season, the land shall yield its produce, and the trees of the field shall yield their fruit."
— Leviticus 26:3-4

During my three-and-a-half years stay in the Brazilian Amazon rain forest, I visited a number of other tribes, including the Banawa, and worked with the Hiberinos, the poor river settlers. I was able to witness the massive destruction of the rain forest not only through slash-and-burn tactics, but also through introducing cattle ranching where it was never intended to be.

According to the United Nations Food and Agriculture Organization and the World Resources Institute, satellite sensing shows that the world's tropical forests are vanishing at the rate of 171,000 square kilometers a year or about thirty-seven city blocks every minute, or two football fields every four seconds. If the current rate of loss continues, all remaining tropical forests, except for a few preserved and vulnerable patches, will be gone within thirty to fifty years.

These forests are the world's key storehouse of biological diversity, an irreplaceable genetic treasure. They provide homes for at least 50 percent of earth's total stock of species. One-quarter of earth's species could be eliminated within fifty years, if current rates of tropical deforestation and degradation continue. Of even more importance are the some 250 million tribal people who live in these forests and depend upon them for their livelihood as hunter-gatherers. All people are made in the image of God, are the pinnacle of His creation and are precious to Him.

Tropical forests supply half of the world's annual harvest of hardwood as well as hundreds of wood products. One-quarter of the world's prescription and nonprescription drugs come from plants growing in tropical rain forests.

Material cycles and energy flow

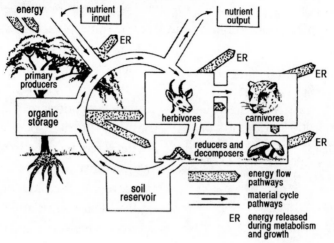

Illustration by Don Richards

Forests also play an important role in the global carbon cycle and as an important defense against projected global warming from an enhanced greenhouse effect. Through the process of photosynthesis, trees help remove carbon dioxide from and add oxygen to the air, explaining why the world's forests are called the "earth's lungs."

True forests are exceptionally complex ecologically and are remarkably rich biotically. The inter-relationship and interdependence of the abiotic and biotic factors in the environment are seen in the extreme in true forests.

Forested watersheds act as giant sponges, slowing down run-off and absorbing and holding water that recharges springs, streams and aquifers. Thus they regulate the flow of water from the mountain highlands to lowlands, to croplands and urban areas and help control soil erosion. This reduces the severity of flooding and the amount of sediment washing into streams, lakes and artificial reservoirs. For example, the repeated ecological disasters in Bangladesh, disastrous flooding and the resulting desertification are mainly due to heavy deforestation in the catchment areas.

Forests also play an important role in local, regional, and global climate. For example, 50-80 percent of the moisture in the air above tropical forests comes from trees by transpiration and evaporation. If large areas of these forests are cleared, annual precipitation decreases, and the regions climate gets hotter and drier so that their nutrient-poor soils are depleted, baked and washed away. Eventually these changes can convert diverse tropical forest into sparse grassland or even desert. I have witnessed this in the Amazon rain forest, where large areas have been cleared of trees to introduce cattle ranching. Regeneration of a tropical forest on such areas may not be possible or, where it can occur, may take hundreds of years.

> Environmental stewardship starts in the heart and in our inner places as we let God do His work in our lives, so we might leave the world a better place from our time on this earth.
>
> — Tri Robinson,
> *Saving God's Green Earth*

Chapter 18: GLOBAL WARMING
AND CLIMATE CHANGE

T here is at the moment huge debate on global warming
— whether or not there is such a thing and, if there is global
warming, what the root causes are. Most people assume that
global warming is caused by burning oil and gas. I believe, however,
that global warming is caused in large part through deforestation.
The Food and Agriculture Organization of the United Nations says
that 25-30 percent of greenhouse gases released into the earth's
atmosphere are the result of deforestation.

Trees are 50 percent carbon. When they are felled or burned,
the CO2 they store escapes back into the atmosphere. According
to FAO figures released in early 2007, some 13 million hectares of
forests worldwide are lost every year, almost entirely in the tropics.
Deforestation remains high in Africa, Latin America and South-
east Asia. I witnessed vast deforestation taking place in the Amazon rain forest during my three years there.

It is also a fact that tropical forest soils are not deep or nutrient
rich. Exchange of nutrients takes place from plant to plant through
their shallow root systems and not through the soil. When slash
and burn takes place, the ash provides sufficient nutrients for a
couple of years and then is depleted altogether, resulting in deserts. *inorganic, lifeless soil*

Delegates of forty-six developing countries at a January 2007
workshop in Rome were greatly concerned by the deforestation
issue. They maintained that 80 percent of deforestation is due to
increasing farmland to feed growing populations. They say, and I
would agree, that the solution is to increase agricultural productivity so that there is less demand to convert forests into farmland.

Here are some thoughts on global warming taken from a presentation that Sir John Houghton gave to the National Association of
Evangelicals in Washington D.C. in March 2005. Sir John, a born-
again believer, is also chairman of the Intergovernmental Panel on
Climate Change (IPCC).

In this presentation, he gave a quick summary of global warming. He said that by absorbing infra-red or heat radiation from the earth's surface, "greenhouse gases" in the atmosphere, such as water vapor and carbon dioxide, act as a blanket over the earth's surface, keeping it warmer than it would otherwise be. This "greenhouse effect" has been known for nearly 200 years and is essential to the current climate to which ecosystems and humans have adapted.

He continues to say that since the beginning of the Industrial Revolution in 1750, carbon dioxide in the atmosphere has increased by more than 30 percent. Analysis of the carbon has demonstrated that this increase is due largely to the burning of fossil fuels such as gas, oil and coal. Action, he emphasizes, needs to be taken immediately to curb these emissions, otherwise the carbon dioxide concentration will rise during the twenty-first century to two to three times its pre-industrial level. If this climate change escalates in the next 100 years, ecosystems and humans will find it very difficult to adapt.

Sir John then considered the impact of climate change or global warming on human communities. He stated that one of the impacts would be a global rise in sea level, because seawater expands as it is heated. This rise would cause the displacement of populations living in low-lying areas. He said a warmer climate would lead to more evaporation of water from the earth's surface, more water vapor in the atmosphere and more precipitation on average. The resulting release of latent heat would trigger more rainfall events but less rainfall in semi-arid areas, resulting in more floods and droughts. Since global emission of carbon dioxide is rising each year, Sir John said emissions from fossil fuels must be reduced. He promoted the development of other sources of energy such as biogas, solar power, hydro, wind, wave tidal and geothermal energy.

Sir John said he is optimistic for three reasons. "First, I have experienced the commitment of the world scientific community in painstakingly and honestly working together to understand the problems and assessing what needs to be done. Secondly, I believe the necessary technology is available for achieving satisfactory solutions. My third reason is that I believe God is committed to His creation. He demonstrated this most eloquently by sending his son Jesus to be part of creation and by giving to us the responsibility of being good stewards of creation. What is more, I believe that we do not do this on our own but in partnership with him — a partnership that is presented so beautifully in the early chapters of Genesis where we read that God walked with Adam and Eve in the garden in the cool of the day."

I include here also excerpts from a briefing by the Evangelical Environmental Network (EEN) on global warming and its effects on Africa:

Africa is the world's poorest continent, with 19 of the world's 25 poorest countries. Nearly 200 million Africans are undernourished, and one-third of African children are stunted or underweight. Sub-Saharan Africa has 10 percent of the world's population, but 24 percent of the world's disease burden. (World Bank 2005)

Because of its poverty, its dependence on locally grown food, recurrent drought and floods, civil unrest and political instability, and diseases like malaria and the AIDS pandemic, parts of Africa are in crisis or live on the edge of crisis. Global warming could make coping with these problems much worse.

Why is this so? Because in Sub-Saharan Africa over 90 percent of agriculture is rain-fed. About 300 million Africans currently live in a water-scarce environment, and by 2025 an increasing population could boost this to 600 million (IPCC Impacts). Approximately 70 percent of African employment is based on agriculture, and agriculture creates over 60 percent of the profits from international trade (World Bank, Washington D.C., 2004). More than half the population lives on food grown locally.

These facts make it clear that any disruption in African's ability to grow rain-fed crops will have serious consequences, especially for the poor.

Other scientists say the belief that emissions cause global warming is a "fairy tale." For example, Dr. Ulrich Berner, a geologist with the Federal Institute for Geosciences in Germany, maintains that global temperatures have varied greatly in the earth's history and

Leading a wilderness trail
in the Umfolozi Wilderness
area in Kwa Zulu
Natal, South Africa

The rumbling Loita Hills,
7000 feet above sea level

A course wilderness trail in the Umfolozi
Wilderness area in Kwa Zulu
Natal, South Africa

A group of Kenyan boyscouts
observing animals in the Ruma National
Park, near Mbita and our base on Lake
Victoria.

The author viewing white
rhinos in the Maasai Mara.
These rhino were
translocated from Umfolozi
Game Reserve in Kwa Zulu
Natal South Africa

Typical African
scavenger, the white
backed vulture,
Umfolozi Game Reserve,
South Africa

Observing elephants whilst leading a
group of adults from overseas on an eco
tour of the Maasai Mara, Kenya

Ugly but beautiful.
A group of warthog drinking
in Umfolozi Game
Reserve

Maasai children dressed traditionally for a wedding, Loita Hills Kenya

A Maasai village in the Loita Hills, Maasailand

Jarawara tribespeople in their village in the Brazilian Amazon

Maasai warriors in traditional dress. These are from left to right: Kashu Parit, David Koyie and Francis Yenko, Andre Brink's co-workers with Walking With Maasai, Loita Hills Kenya

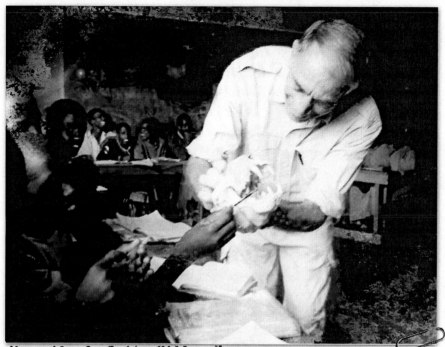

Maasailand, Loita Hills, Kenya..
Don teaching adaptation to Maasai school children,
holding a leopard skull

Environment and Resource
Stewardship school in Mbita,
researching water quality on Lake Victoria

Don showing a group of
Zulu youth on a wilderness
trail at Lake St. Lucia
in South Africa, checking
out hippo footprints

are unrelated to human activity. He also states that carbon dioxide does not cause climate change — that climate changes have always occurred and will do so in the future. Scientists who agree with him attribute climate change to factors such as our planet's orbital eccentricities, axial wobble, solar brightness variation and cosmic rays, and plausible terrestrial drivers such as volcanic activity and tectonic movements. (JuncScience.com, April, 2006)

There is clear evidence that at one time far before the Industrial Revolution and invention of the internal combustion engine, tropical plants grew at the North Pole. At the end of the Pleistocene era, the ice that had covered much of North America, Europe, and Asia began to melt. Quite a history of temperature range has taken place on this planet, and I believe that scientists are not fully cognizant of all the factors involved.

> The initial step for a soul to come to knowledge of God is contemplation of nature.
> — Irenaeus of Lyons, 125-202 AD

I present here another view by Dr. E. Calvin Beisner, who is the associate professor of historical theology and social ethics at Knox Theological Seminary. He is a renowned expert on environmental issues and testified in 2006 before a U.S. Senate committee on topics related to the environment. He is also author of several books, including Where Garden Meets Wilderness: Evangelical Entry into the Environmental Debate. He has this to say regarding the global warming and climate change debate in a book authored by him and Dr. James Kennedy entitled Overheated: A Reasoned Look at the Global Warming Debate.

Dr. Beisner replies to Dr. Kennedy's request to explain global warming and climate change in layman's terms with an explanation of how concerned we should be. "Historically the globe has warmed and cooled in cycles throughout its history. The current, or recent, warming has not been outside the range of natural variability."

"What alarms many people is the idea that human action – particularly in emitting carbon dioxide from the burning of fossil fuels

— is the cause of the recent warming. However, there really isn't any good scientific evidence supporting this idea, because most of the warming took place before the increase of CO2. There was actually a significant cooling period from the mid-1940s to the late 1970s, while CO2 was increasing rapidly, and there was another increase from 1979 to 1998. In fact, there has been no warming since 1998, and even a slight statistically insignificant cooling, despite the fact that CO2 has continued to rise. There is not good strong proof that CO2 is driving global warming.

"What is settled science is that the globe has warmed recently. What is not settled science is that human activity causes it. What is also not settled is that global warming is going to go beyond the bounds of natural variability and is going to be harmful to human beings and the rest of the biosphere. ... history tells us that the Earth's warmer periods have been healthier for human beings than the colder periods ... the Medieval warm period that ran roughly from about 950A.D. to about 1350A.D. or 1400 A.D. was warmer than the most recent period in the late twentieth century, which was also the period of best harvests, of the fewest major storms, and a real thriving of the human population.

"The Little Ice Age that began around 1450 A.D. and ended around 1850 A.D. was the period of the worst harvests and greatest human suffering in the Northern Hemisphere. Therefore, our history tells us that warmer is better for us, not worse. ... historically change is normal for the Earth. A static temperature for the Earth has never been the case.

"It is actually very difficult to try to allocate the causes of global warming. Most climatologists will say it really can't be done. Nevertheless, some people who are trying to promote the policy of reducing CO2 emissions have claimed that human action has been the major cause of global warming. Their claim is not credible for a number of different reasons. First of all, the correlation doesn't fit between CO2 and temperature. Second, the correlation is very strong between temperature and solar output — solar output of two sorts: One, radiant heat energy, and that changes in cycles and, two, output of what is called 'solar wind' that also cycles up and down.

"In October of 2006, a study by the Danish Space Center was released proving that as solar wind increases, it blocks more cosmic rays which cause the formation of cloud droplets in our lower atmosphere. The more clouds there are, the more heat energy gets reflected back into space, rather than warming the Earth.

"The more solar wind there is, the fewer clouds there are, and therefore, the warmer the temperature. The less solar wind there is, the more clouds there are, and the colder the Earth. ... My own estimate is that human contribution is minute — certainly not significant.

"Another reason why it isn't safe to say that human activity is a major cause of global warming is because the correlation is very strong between sun cycles and global temperature, but it is not strong between CO_2 emissions and global temperature."

"More important is the cycling in solar wind. Solar wind blocks cosmic radiation from coming into the Earth's atmosphere. The more cosmic rays come in, the more low-level clouds there are, and low-level clouds reflect solar heat energy back into space. So if there are more clouds, the Earth is cooler; and if there are fewer clouds, the earth is warmer. As solar wind cycles up and down, cosmic rays cycle up and down in reverse, and clouds cycle up and down. Consequently, solar wind is very important to Earth's temperatures. That is the best correlation we have, and surely the cause of most of the warming of the twentieth century. That was the conclusion of the Danish Space Center study.

"The greatest danger that is predicted from Global Warming is an increase in sea level. And yet, the data of the Intergovernmental Panel on Climate Change, which is the large United Nations body that studies these things, indicates that sea level rise over the next century will be perhaps 4 to 15 inches total—less than 1.5 inches per decade. This rise is a rate we can readily live with and readily adapt to at much lower cost than we could if we tried to prevent reducing human CO2 emissions.

"Another danger, which is discussed, is more frequent storms, but climate history tells us that storms are more frequent and more intense in colder periods than they are in warmer periods. Also feared is a loss of agricultural production, but climate history tells us there are better crop yields in warmer years than in colder years. In fact, CO2, which I believe people improperly blame for global warming, is actually a fertilizer for plants. They use it in photosynthesis. Actually, for every doubling in CO2, there is a 35 percent increase in plant growth efficiency, which also gives an increase in crop yield."

In June 2007, a program written by Brian Fagan, entitled "Little Ice Age — Big Chill" was featured on the History Channel, agreeing with the majority of what Dr. Beisner had to say. It demonstrated very clearly the effects of the temperature change on history during the Little Ice Age and the medieval warming period.

In declaring His covenant with all creation, God says to Noah: *"While the earth remains, seedtime and harvest, cold and heat, winter and summer, and day and night shall not cease."* (Genesis 8:22)

We know that God is faithful to His covenant promises.

From a Christian perspective, we cannot be alarmists. Psalm 103:19 states, *"The Lord has established His throne in the heavens and His sovereignty rules over the universe."* Psalm 104 and many other Scriptures are clear that God is in control over weather, earthquakes and all life.

In conclusion, there are huge differences in thought about global warming and climate change and the probable effects on the planet. Man's predictions are as sure to be correct as tonight's weather forecast!

However, as to global heating, in the reference to the day of the Lord, which will come like a thief in the night, in which the heavens will pass away with a great noise, and the elements will melt with fervent heat, both the earth and the works that are in it will be burned up. (2 Peter 3:10)

Chapter 19: TROPICAL RAIN FOREST SUCCESSION

"Then God said, 'Let the earth bring forth grass, the herb that yields seed, and the fruit tree that yields fruit according to its kind, whose seed is in itself, on the earth'; and it was so. And the earth brought forth grass, the herb that yields seed according to its kind, and the tree that yields fruit, whose seed is in itself according to its kind. And God saw that it was good."

— Genesis 1: 11-12

Succession is a process that involves changes in plant communities over time. For example, cleared areas over time naturally become covered by plant communities that replace or succeed each other. This sequence of communities is called the sere, and the transitory communities are called seral stages, developmental stages or pioneer stages. The terminal stabilized system is the climax.

Grass is one of the first stages in an area being restored to its former self, followed by secondary succession of richer grasses, then bushes, and then trees. Important trends in successional development involve greater storage of materials and increased cycling of major nutrients such as nitrogen, phosphorus and calcium.

In tropical rain forests the orderly successional stages are not that clear. Rather, change is due to the refilling of gaps in the vegetation made by the collapse of canopy trees. Species that thrive in newly formed gaps are waiting dormant until conditions are right. Pioneers (the first plants to move into a degraded area) are not found in the under storey of mature forest, but their seeds germinate in gaps open to the sky and sunlight, because these plants cannot survive in the shade.

Pioneers also have most of the following characteristics:
■ They have small seeds, produced in large numbers that are dispersed by animals such as insects and armadillos or wind
■ Their dormant seeds are usually abundant in forest soil
■ Their height growth is rapid

- Branching is relatively sparse, and leaves are short-lived
- Rooting is superficial
- Wood is pale and of low density
- Leaves are usually large and are eaten by herbivores and insects

Some pioneers in the jungle persist into the canopy and become large trees. Frogs, toads, ticks and mosquitoes are more common in gaps and in secondary forest, where the herbaceous layer in which many animals live is near the ground. Gap plants tend to have brightly colored flowers that are visited by bees and birds, contributing to cross-pollination.

A true forest is made up of a number of layers.

The amount of sunlight that reaches the ground cover in a true forest has been measured as one six-hundredth of the sunlight hitting the canopy. Thus the lower level plants tend to have larger leaves than those in the canopy to compensate in the process of photosynthesis.

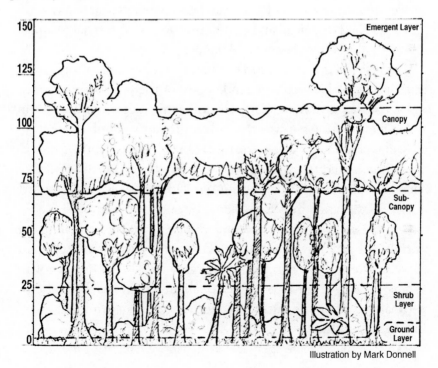

Illustration by Mark Donnell

Chapter 20: **SURVIVAL**

"And this is the victory that has overcome the world – our faith. Who is he who overcomes the world, but he who believes that Jesus is the Son of God?"

— 1 John 5:4-5

I was co-leading a YWAM Community Development School at a survival camp in the Amazon. Mark Barnes, the school leader, Daniel, a YWAM missionary working with the Banawa, and Bidu, a born-again Banawa chief made up the rest of the leadership.

We had hired the help of local residents and two of their engine-driven canoes to transport us up the Rio Macoin to our starting point. It was a day-long haul, chugging slowly upstream. Thick overhanging canopy covered the river in its narrow stretches, but every now and again we would enter into bright sunlight as the river widened. Brightly colored macaws and green parrots squawked and screeched overhead and caymans sunbathed on muddy banks giving us glimpses of beauty and danger.

Our first night was spent in our hammocks strung up between trees in a small hiberena hacienda built for the purpose of processing and cooking manioc. Early the following morning we set off in single file, hacking our way through nearly impenetrable jungle and crossing a number of clear streams that trickled out of the forest floor. The humidity was unbearable. Soaked in so much sweat that we would often have to wring out our T-shirts, we were persistently attacked by myriad insects attracted by our perspiration. These stream crossings offered a welcome respite as we lay in them to cool ourselves. Every now and then snakes would slither across our path as we progressed through thick undergrowth. Thick vines, known as Sipo de Agwe, cut into lengths of about a meter each and held vertically above our mouths, produced good clean drinking water to relieve our thirst.

Our two nights were spent cooking fish we caught in the streams and roasting a monkey we had shot during the day. A tough meal! Hammocks strung between trees and cut palm branches provided a makeshift shelter from tropical rain during the night.

Our return trip down river to civilization became a horrendous affair. The first canoe with half our group went ahead and arrived without a problem. As we set off in the second canoe, our problems were only beginning. First, we noticed our driver drinking pure alcohol. After about two hours we developed propeller shaft problems. Pulling to shore, we disassembled the motor and spent the next three hours repairing the engine and shaft. By this time our driver and assistant were barely coherent as a result of their alcohol consumption. We tried to persuade them to hand over driving to either Daniel or Bidu, both excellent river pilots, but they flatly refused. Our canoe's course became very erratic, zigzagging this way and that across the river, bumping into the banks and hitting logs and trees.

> Hans Nielsen Hauge (1771-1824) preached that all we own and all that we are come from God. The Lord expects us to be good stewards of what He has given us.
>
> — Loren Cunningham, *The Book that Transforms Nations*

Soon it became dark and difficult to see tree trunks that had fallen across the narrow river from bank to bank. Cayman eyes were reflected now and again in the dim torchlight. Much prayer went up to heaven that night, and as usual God was faithful, enabling us to reach our destination, albeit five hours late.

I began this chapter by quoting 1 John 5 about how faith in Jesus Christ overcomes the world. I believe that faith in Him brought us through that night in a perilous situation that could have turned into tragedy. Faith always triumphs in trouble.

Paul speaks in Romans 5:1-4: "*Therefore, having been justified by faith, we have peace with God through our Lord Jesus Christ, through whom also we have access by faith into this grace in which we stand,*

and rejoice in hope of the glory of God. And not only that, but we also glory intribulations, knowing that tribulation produces perseverance; and perseverance, character; and character, hope."

Whilst writing this chapter in our cabin in the Great Smoky Mountains, I watched a number of squirrels leaping from treetop to treetop some eighty feet up without any hesitation at all. My thoughts were that although squirrels probably do not know anything about faith, we can learn so much from them. The Bible speaks of this in Job.

"But now ask the beasts, and they will teach you;
And the birds of the air, and they will tell you;
Or speak to the earth, and it will teach you;
And the fish of the sea will explain to you;
Who among all these does not know
That the hand of the Lord has done this,
In whose hand is the life of every living thing,
And the breath of all mankind?"

— Job 12: 7-10

Why is it necessary for missionaries to undergo survival courses? Missionaries going into developing countries are going to meet rigorous conditions that will test their ability to survive. Missionaries going out to unreached people groups will be even more stretched. They will be faced with tough situations: Unknown travel means, strange food, lack of facilities and extreme hardships. As we go out into missions, we are forced to come out of our comfort zones.

Jesus tells us in Mark 10:29-30, "Assuredly, I say to you, there is no one who has left house or brothers or sisters or father or mother or wife or children or lands, for My sake and the gospel's, who shall not receive a hundredfold now in this time — houses and brothers and sisters and mothers and children and lands, with persecutions — and in the age to come, eternal life."

If we are to spread the gospel of Jesus Christ and be His ambassadors in extending the Kingdom of God on all the earth, we will need to make sacrifices and leave our comfort zones. There is a very true saying that when the conditions become tough, the tough keep going!

Chapter 21: **MOVING ON**

The year 1994 saw more changes in my life. In June, a YWAM Kings Kids team known as "HANDS" (Helping A Neighbor Develop and Serve), led by Gail and Mike Kent from Kona, Hawaii, arrived at the Porto Velho base in the Amazon to help build the Environment and Resource Center on the base. This was in preparation for the first YWAM EARS school due to start in mid July. Amongst the HANDS staff was a beautiful lady named Nancy Donnell from the U.S.A., to whom I was immediately attracted. Little did I know at first that she felt the same way about me. To cap it all, the Lord impressed on me during my quiet time that she was the one to be my wife. The Lord also spoke to her. We hardly spoke to each other in the business of construction and ministry. We had both been married before, and both of us were called by the Lord into missions, Nancy to work with orphans and I in the area of environmental stewardship.

> The hour is late, the need is urgent; God is stirring His people to break down dividing walls so that we might work together in unity to "tend his garden."
>
> — Tri Robinson, *Saving God's Green Earth*

Nancy left with the HANDS team after six weeks of service with us, and I plunged immediately into leading the first EARS school from mid-July till the end of November. During this five-month school, God clearly directed me to return to Africa, to start an EARS ministry in Kenya. Little did I know that the Lord had also called Nancy to Africa to work with orphans. We corresponded with each other and, to cut a long story short, we were married in March 1995. Together with Andre Brink, one of the EARS students, we launched "Kenya Islands Mission" in July 1995. This ministry was later to become YWAM, Kenya Islands Base, on the shores of Lake Victoria.

Through God's word to us from Isaiah 41:1, "Keep silence before

Me, O coastlands, and let the people renew their strength!" and through circumstances, Nancy, Andre and I were drawn to Lake Victoria. We bought a piece of land on the lakeshore close to the small town of Mbita. This rural town is in the Suba District of the Nyanza Province of Kenya. The three-and-a-half acres was land covered in short scrub and sloped down to the lakeshore. Land had to be cleared for the first building, which was to be our home, office and quarters for some staff.

As we entered the district we noticed the devastation from excessive logging, overstocking of cattle and goats and the erosion and lack of clean water that resulted. We learned that the Suba District was the poorest in Kenya and had the highest incidence of AIDS (close to 42 percent positive), with a large number of orphans as a result. The area was rife with witchcraft and demonic worship, including "night runners" who ran around naked at night putting curses on homes and people.

God says in Hosea 4:6, "*My people are destroyed for lack of knowledge.*" And in Jeremiah 9:12b and 13a, "*Why does the land perish and burn up like a wilderness, so that no one can pass through? ... because they have forsaken My Law which I set before them.*"

Human sin and neglect have a tremendous impact on the whole of creation, which in turn affects those who live in it. Yet when we turn back to God and repent, the creation is restored.

"It shall come to pass in that day,
That I will answer," says the Lord;
"I will answer the heavens,
And they shall answer the earth.
The earth shall answer,
With grain,
With new wine,
And with oil;
They shall answer Jezreel.
Then I will sow her for Myself in the earth,

And I will have mercy on her who had not obtained mercy;
Then I will say to those who were not my people,
You are my people,
And they shall say, 'You are my God!'"

— Hosea 2:21-23

Hence there is an urgent need for the people of the world to turn back to their Creator and submit to His laws.

This is what YWAM Kenya Islands Base, ministering to the shores and islands of Lake Victoria, is trying to achieve through the gospel of Jesus Christ, discipleship and education. It teaches about God's love for all creation, stewardship of that creation and the resulting benefits for humankind.

Nancy and I felt this call to relief and restoration that is borne through the grace of God and out of Him who is father of the fatherless. (Psalm 68:5) He desires to multiply the fruit of the trees and the increase of the fields, so that people may never again bear the reproach of famine. (Ezekiel 36:30) YWAM Kenya Islands Base is reaching out to hurting communities on the 14 islands and along the shore through its Bible-based community development, environmental education, resource stewardship programs and discipleship training. It is enabling people both to have a personal relationship with God and to look after their own needs.

Chapter 22: **CREATION AND RESOURCE DEVELOPMENT**

T he Bible is full of instruction concerning the environment, land, work, plants and animals. In creating the world, God first made an environment, and then placed Adam and Eve into it. Following this pattern, let us look at the environment of creation and then examine humanity's role.

Several details indicate that the Garden of Eden was a sanctuary. It was a place where God was present in glory. (Genesis 3:8) In biblical times the sanctuary was specially designated as a place of worship, and we still consider it so. But as believers in Christ, we believe that we should worship God in everything we do.

Romans 12:1-2 says, *"I beseech you therefore, brethren, by the mercies of God, that you present your bodies a living sacrifice, holy, acceptable to God, which is your reasonable service. And do not be conformed to this world, but be transformed by the renewing of your mind, that you may prove what is that good and acceptable and perfect will of God."*

Then in 1 Corinthians 10:31, Paul tells us, *"Therefore, whether you eat or drink, or whatever you do, do all to the glory of God."*

In the Bible's hierarchy of environments, the sanctuary holds first place. On the first day of his life, Adam was placed in the garden where he was to worship and walk and talk with God. This was a model of the earthly sanctuary, a sacred and safe place. Though our homes are not a sanctuary, they become something like a sanctuary, a place where we feel safe and meet with God. The biblical ideal is for every man to have his own garden, his own sanctuary. (Micah 4:4) I believe that the world today is an analogy to the garden, a place that should be holy and sacred and safe.

Let us turn to the creation mandate given to Adam and Eve. In Genesis 1:28, God issues this command: *"Be fruitful, multiply, and fill the earth, and subdue it, rule over the fish of the sea, and over the birds of the air, and over every living thing that moves on the earth."*

In five imperatives, the Lord outlined His program for man's labors in the world, and the passage provides material for Christian reflection on the environment. For example, this verse challenges contemporary assumptions about population growth and control. Certainly since creation there has been population growth, which today has been termed a population explosion. Every two seconds five people are added to the world population. During the period as hunter-gatherers, the human population was about 5 million. In 8000 B.C., mankind became farmers, and populations were about 100 million. At the time of Christ it rose to about 169 million and over the next 1,650 years the population grew to 500 million. The first billion came in about 1830; the second billion in 1930. It took only thirty years to reach 3 billion in 1960, and fifteen years to reach 4 billion in 1975. As of 2006 the figure stood at 6.5 billion. (CIA Fact Book 2006)

One theory of the cause of population explosion is improved medical resources and practices, resulting in longer life spans. A recent theory is that growth of population stems from social insecurity during urbanization. It has been noted that during times of war populations grow, a sort of survival technique to assure that families are preserved. We see this trend especially in developing countries with widespread poverty and rapid urbanization.

> Nearly one third of the world's cropland has been abandoned in the past 40 years because erosion has made it unproductive.
>
> — **www.unep.org**, 2006

The world looks at birth control as the answer, but the Bible has another view. My thoughts are that the obvious problem lies in where people have settled. If we study settlement geography, we observe that most cities have developed where large rivers meet or enter the sea, at good natural harbors, in wide valleys, where trade routes and roads cross, and below mountain passes. This leads to over-population in those areas, overtaxed resources, pollution, soil erosion and millions of people without proper sanitation, health care or access to safe water. Millions of people flock to cities where they believe life will be easier and jobs will be available. These desires are often dashed and the result is the formation of huge slums.

The 20th century witnessed the rapid urbanization of the world's population. The global proportion of urban population increased from a mere 13 percent in 1900 to 29 percent in 1950 and, according to the 2005 Revision of World Urbanization Prospects, reached 49 percent in 2005. This trend also leads to less agriculture, and so maybe the answer lies in people returning to the land and becoming producers of food. So I believe that it is not a question of population explosion but of population distribution.

The Christian response to all this is that we should take active part in city and community planning, new settlements, and new areas for farming and industry. We should also be engaged in helping to improve living and working conditions. Social and economic tasks should go hand in hand with the redemptive power of the gospel, which is to be proclaimed and lived out among the needy. *"The Spirit of the Lord is upon Me, Because He has anointed Me to preach the gospel to the poor."* (Luke 4:18a) Adam and Eve were commanded to bring the earth, its resources and other creatures into service to mankind and ultimately to promote the greater glory of God. To rule or take dominion applies to Adam and Eve's relationship to the rest of creation: To lead, to take care of and not to trample.

We have all been made in the image of God. God in His infinite wisdom meant mankind to have the best out of life. We know that sin has tarnished that, but as Christians we are to be creative on a finite level and should certainly have more understanding of our responsibility to treat God's creation with sensitivity. We should be developing our talents to do something about it. As Adam and his descendants fulfilled the calling to subdue the earth and rule the lower creation, the entire world was meant to become more and more like a sanctuary. God's intention was that human action was to make the world more productive, more secure and more beautiful, a sanctuary worth guarding. However through sin coming into the world, human beings rule and subdue the earth, not in righteousness and for the glory of God, but in rebellion. Instead of transforming creation into an image of the heavenly Jerusalem, they erect Babel. However, by union to the last Adam, becoming more and more like Christ, men and women are "put back on track," restored to their Adamic calling, influencing the world to become more like a sanctuary.

Now as we look back at all this, let us look at what has gone wrong in our role of making the earth a better place to live and in developing it for good. I once viewed a documentary entitled, Because of lack of interest, tomorrow has been cancelled! It made a great impression on me as I watched how our lack of interest in things environmental, involved us in the business of polluting and raping the earth and how this is leading to so many tomorrows being cancelled for so many millions of people throughout the world. To correct this, I will point out various areas that are vital and need our urgent attention.

UNNECESSARY DEATHS

If we look at unnecessary deaths, we can see that one-quarter of a million children, mostly under the age of five, die each week from frequent infections and prolonged under nourishment. Of those who do not die, many will have ill health and poor growth, restricting full mental and physical development.

It is possible to prevent most childhood deaths and malnutrition through basic health care and nutrition programs. It has been estimated that 14,000 young lives can be saved each week by informing and supporting parents in basic and inexpensive actions, such as immunizing. If fully applied, such programs could reduce infant mortality by 125,000 per week. Of the 14 million childhood deaths each year, most are caused by preventable and curable diseases such as diarrhea, measles, tetanus, whooping cough and respiratory infections. Improved education on diseases and good communication would also be a huge asset. In many rural areas in developing countries, the lack of transport and clinics as well as trained medical staff, together with horrendous road systems, make it impossible for the sick to get to medical help.

Health is not the absence of disease, but a well-being of body, soul and spirit. High mortality rates are usually related to extreme poverty. In developing countries, three-fourths of all deaths occur between the ages of 45 and 65. In the Suba District of Kenya where we have been working, the latest official life expectancy figure is 37 years! Life expectancy in most African countries is in the 30s and 40s. The high mortality rates in the developing world are similar to those in Europe some 150 years ago. Diet and lifestyle are the root causes of illness and death. In poor countries, parents see their children as hope for economic support in the future; thus, because of the high death rate, there is a high birth rate. We have witnessed widows and children stooping to prostitution in the Lake Victoria area, prevalent in the some 121 fishing villages in the Suba District where so many of the fishermen have very low moral standards and are away from their own families and home for most of the year.

> We cannot be excused when we have not at all considered God in His works. He does not at all leave Himself without witness here – Let us then only open our eyes and we will have enough arguments for the grandeur of God, so that we may learn to honor Him as He deserves.
>
> — John Calvin, 1509-1564

These fishermen practice a high-risk tradition known locally as "Jaboya." This practice requires a woman who wants to be a primary fish buyer in order to make a living by selling fish at market to have sexual relations with the fishermen staying at the beach. Another practice is one called "Abila." Here boat owners, who have crew houses on the beaches, hire women to cook and clean for them in exchange for sex with them and their crews. This happens with children also, when the fishermen force them into prostitution in exchange for work. All this has led to the Suba District having the highest HIV prevalence in Kenya. This is one of the target areas of YWAM in Mbita, where we are try-ing to evangelize, disciple and teach these fishing communities moral standards and environmental stewardship of the lake. This is just an example of what is taking place in many parts of the world.

THE FUTURE GENERATION

Because the majority of the world's populations ignore children and teenagers, the enemy is using every possible strategy and delu-sion to fill the spiritual void in these generations. The void is there because the truth concerning God is not being communicated to them. This is especially so in most of our education systems. There is a God-given spiritual capacity within children and youth that is not being taken into consideration. The world's youth are our future leaders, and they have great potential. Fortunately we see a great shift today, with thousands of youth on the march for Jesus through such organizations as The Call, The International House of Prayer, YWAM, Campus Crusade, Youth for Christ and others. However, there is still a great need to disciple millions of young people worldwide.

That there is a veritable war against children is evident in the issues of abortion and child pornography. Where are the Esthers, Davids and Gideons of our day? They are waiting to be exposed to the truth. Dutch Sheets, in his book Praying For America, relates some alarming statistics about what is happening to our youth in America: "Over 2,000 teenagers commit suicide each year — six each day; 3,610 teenagers are assaulted and 80 raped every day. That's 1,317,650 assaults and 29,200 rapes annually; 1,106 teenage girls get abortions every day — 403,690 each year; 4,219 teens per day contract sexually transmitted diseases (1.5 million each year). One can see the ploys of Satan trying to wipe out a generation."

THE ELDERLY

The challenge of missions to the elderly and education of the elderly must not be overlooked. John 10:10 says, "The thief does not come except to steal, and to kill, and to destroy. I have come that they may have life, and that they may have it more abundantly." The elderly need to have better care and to be prepared for "everlasting life."

"Listen to Me, O house of Jacob, and all the remnant of the house of Israel, who have been upheld by Me from birth, who have been carried from the womb: Even to your old age, I am He, and even to gray hairs I will carry you! I have made, and I will bear; Even I will carry, and will deliver you." (Isaiah 46:3-4)

> Deforestation affects the amount of water in the soil and groundwater and the moisture in the atmosphere.
> — http://wikipedia.org

Here it shows the Lord's never-ceasing care. We need to emulate the Lord in the care of the elderly, and programs for the elderly must not be overlooked in our planning.

THE FAMILY

The family is the first institution made by God, which developed into tribes and then nations. A close and caring family is important in the development of healthy children, and the bonding in the family so important to the mental and emotional health of each child into adulthood. Parenting should take the form of tutorship based on the Hebrew education system and not be left to the Greek-based form of education at most schools. Children learn more in the first five years than during the remainder of their lives. A child's personality is established by the age of twelve. Thus the family should be the primary source of education, security and support.

In these days, the family institution is under attack through high divorce rates, abuse and lack of parental attention. As we work with cultures and environmental education programs, we need to understand and work with family units and to have programs for preschool and younger children.

Here follow some Scriptures that relate to the importance of children and how they should be raised in families:

"At that time, Jesus answered and said, 'I thank you, Father, Lord of heaven and earth, that you have hidden these things from the wise and prudent and have revealed them to babes.'" (Matthew 11:25)

"And Jesus said to them, 'Yes. Have you never read, Out of the mouths of babes and nursing infants You have perfected praise?'" (Matthew 21:16)

"Train up a child in the way he should go, And when he is old he will not depart from it." (Proverbs 22:6)

CREATING JOB OPPORTUNITIES

In light of the way women sell themselves in the fishing villages for money to survive and how widows even send their children into the villages as prostitutes, we see how important it is to create job opportunities in the form of small businesses. This strategy applies to men, as well, since a man without work is prone to feel that he has no worth.

Creating job opportunities is an integral part of environmental education and community development programs. One of the ministries that YWAM Mbita in the Suba District of Kenya facilitated is called "Bananas of Hope." We got to know Margaret Nyange, a widow who had a great talent for making beautiful baskets out of banana fibre. As we encouraged her, she gathered a number of widows around her and taught them how to make the baskets. She would lead them in daily Bible studies. The Lord enabled us to find an overseas market for their produce, and now containers of these baskets are being shipped to the U.S.A. These widows are all earning a wage and having hope for the future.

Christopher Opiyo, one of our leadership team at YWAM near Mbita, has been working with 10 self-help groups of fishermen, who are now into market gardening, growing food not only for themselves, but also as a source of income during the non-fishing season. Banding together, these groups have made enough to buy and share a water pump to irrigate their crops.

Developing countries will have to solve the challenges of sustainable agriculture and food production before they can enter and industrial stage. There is a need for more agro-intensive programs to create meaningful jobs and self-employment to generate income to raise their quality of life. Unemployment and underemployment will probably be the greatest causes of social and political unrest in the years to come.

SUSTAINABLE FOOD PRODUCTION

There is promising agricultural development in almost every region of the world. The world produces more food per head today than ever before. Even so, over 850 million people do not receive enough food to lead a fully productive life. (World Health Organization)

Three main types of agriculture exist today, and these are: (a) Industrial agriculture in the first world; (b) Green revolution agriculture in resource-rich, flat, irrigated areas; (c) Resource-poor agriculture relying on rain. This type of agriculture is found in the majority of dry lands, islands, and forests with poor, fragile soils. People need to be taught alternative agricultural techniques, and food needs to be produced where they live, so that families can support themselves. Although food aid is a worthy cause, it has the danger of reducing incentive and altering the economy. This does not apply, of course, to the first steps in relief work.

SOIL

The protection and wise use of soil is of utmost importance. God placed people on earth to be fruitful, multiply, fill the earth and rule over it. (Genesis 1:28) Because of mankind's sin, greed and negligence, God's plan was never properly implemented. In most cases, land has been selfishly used. Soil, once considered an unending resource, is threatened by mismanagement and ignorance. The best food producing lands have suffered extreme losses through soil erosion. Only 85 million square kilometers of the earth's 149 million square kilometers is reasonably biologically productive. (*Topsoil Loss — A Global Perspective*, by Bruce Sundquist)

This neglect is usually offset with vast amounts of fertilizers that are expensive and cause problems of their own.

Agricultural land degradation is primarily a result of desertification and drought, both of which can be caused by humans, especially through deforestation and overgrazing. This is evident on almost every continent. For example, in 2005 soil erosion exceeded soil formation in one-third of U.S. cropland. In the U.S.A and Europe, erosion rates are 1,700 tons per square kilometer annually. In Canada, soil degradation has cost farmers $1 billion a year. India's soil erosion affects 25-30 percent of total cultivated lands. Asia, Africa and Latin America together stand to lose some 800 million hectares of rain-fed cropland if corrective action is not taken. In South Africa alone, almost 400 million tons of topsoil is washed into the sea each year. Africa leads in soil loss, exceeding 40,000 tons per square kilometer a year! More than 40 percent of agricultural land globally has degraded soil.

> If all the world's water were fit into a gallon jug, the fresh water available for us to use would equal only about one tablespoon.
> — **www.lenntech.com/water**, 2008

The growing numbers of floods and droughts around the world have been linked to deforestation, which has occurred on 29 percent of the earth's land surface. More than 850 million people live in these regions. Six percent of the earth's land is classified as severely desertified. Deserts are growing at an annual rate of 6 million hectares. All this presents a dismal picture. God called us to take care of the earth: "Let them rule — over all the earth." (Genesis 1:26) It would be well to heed the warning found in Revelation 11:18, which foretells a coming judgment in which God will destroy all those who destroy the earth.

WATER

Water is essential to life. It accounts for 50-75 percent of a human's body weight, 90 percent of our blood plasma, 75 percent of man's muscle tissue and 25 percent of dry bone. The human body can survive only a few days without water. Yet about 2 billion of the world's inhabitants depend on contaminated water, which is believed to cause between 25 and 30 million deaths a year, about 60 percent being of children under the age of 5.

Untreated and reused water in some developing nations carries germs that cause around 80 percent of diseases. Typhoid, cholera, dysentery and schistosomiasis are associated with contaminated water, causing 2,001,000 deaths (WHO 2002) each year. Diseases also are responsible for loss of productivity, economic instability and poor quality of life.

The water crisis requires understanding, wisdom and committed action by individuals and groups. Each region has unique problems that must be studied. The inhabitants must be made aware of water contamination and its sources and how to make corrective action. Economical and effective water treatment must be developed and implemented where natural purification is unreliable. Wastewater recycling systems need to be introduced where water is scarce. Water conservation is an imperative.

Channeling clean water to replace unsuitable water can also be a channel to relate the Good News. When Jesus asked us to give water to the thirsty, He meant pure healthy water, not contaminated, disease-causing liquid!

EDUCATION AND LITERACY

It is evident that education is part and parcel of an environmental education program in any community. Environmental education across a school curriculum based on a biblical view of God and man as well as of the world can be a very effective tool.

Hand in hand with education comes literacy. Understanding literacy levels and the kind of education in an area is important for planning effective evangelism and discipleship and for improving people's well-being, self respect and the quality of life. I have witnessed areas in the world where governments have forbidden ethnic tribes to use their own language and have limited their education. This leads to a sense of inferiority and opens them to manipulation by the unscrupulous.

David Koyie, a Maasai working with Andre Brink, has pioneered an effective literacy program amongst Maasai warriors in the Loita Hills. It is wonderful to see these warriors, bedecked in their traditional warrior costume complete with spears, sitting in a classroom and having the world of reading and writing opened up to them. Literacy not only has a major effect on production and economics but also is essential in enabling people to read the Word of God. Bible translation ministries such as Wycliffe, S.I.M., YWAM and others are doing great work in this area.

> Most of the world's people must walk at least 3 hours to fetch water.
>
> www.lenntech.com/water, 2008

There are some 800 million illiterate adults in the world, mostly in developing nations. The majority are women. We need to be involved in supporting this work.

ENERGY

In the Western world we take energy much for granted. Electricity and gas are all very much part of our lives, but we need to compare our consumption of the world's energy resources to that in developing countries. The imbalance of world energy consumption is reflected in the more than eighty times greater consumption per person in industrial market economies than in sub-Saharan Africa. One-quarter of the world's population consumes three-fourths of the world's primary energy.

We need to discover and develop energy-efficient technologies that can be low-cost and simple. I realize that work is being done in that direction, but most of the alternatives are too costly for third world nations. Solar energy is a wonderful alternative, but the initial cost of purchase and installation is prohibitive to people in the developing nations. The same applies to electric driven vehicles and other avenues of change in the energy field. Seventy percent of the people in developing countries use wood as a primary source of energy. They burn about 700 kilograms of wood per person annually. Around 24 billion people are living where wood is scarce.

In China, 17 percent of energy resources are used just to boil water. Population growth and migration to cities give governments of developing countries no option but to organize their agriculture to produce large quantities of wood and other plant fuels. Fuel wood scarcity and deforestation are not the same. Industries in cities cut down forests to meet the needs of urban dwellers. The rural poor do cut down trees and do not replace them, but most of the poor usually collect dry branches that are lying around. In many areas of the world where there is no longer wood for fuel, cow dung, crop remains and husks are used, which robs the soil of much needed nutrients.

Another result of fuel shortages in these areas is malnutrition, not just through lack of cooking fuel but also, as noted, through erosion, infertile soil and the resulting poor crops. There is a great need for agro-forestry programs in many of these areas, as well as initiatives to produce low-cost hydro-power, solar power or harnessing the wind.

We can see that stewardship must be brought about through individual responsibility, resulting in sensitive conservation, creative innovation and deliberate implementation. Unfortunately, the substance abuse that seems so prevalent among the poor, many of whom see no hope for the future, produces negligence in these areas. Alcohol and drug abuse are barriers to solutions, but we know that Jesus Christ can fully restore substance abusers to health and that He is the hope now and in the future.

POLLUTION

Let us again return to the analogy of this world and the sanctuary. The Bible speaks frequently about clean and unclean, abominable things and pollution. In the Bible, pollution has mainly to do with transgressions of God's laws and particularly laws governing man's approach to God's sanctuary. (Deuteronomy 23:14)

For a polluted man or nation, the sanctuary took on qualities opposite to the qualities of the original undefiled garden. After Adam was polluted by sin, the sanctuary was no longer a refuge. Instead it was a dangerous place with cherubim stationed at the entrance to prevent return. (Genesis 3:24) It was no longer a place Adam could flee to but a place he had to flee from. A polluted sanctuary provides no refuge.

Nor was the garden a place where defiled Adam could be nourished. He was denied access to the tree of life. Thus the sanctuary was no longer a fruitful and productive environment for him. (Deuteronomy 28:23) So we can see that sin disfigures the sanctuary and that the connection between sanctuary and environmental pollution is that sanctuary desecration pollutes the world.

Man is therefore responsible for environmental pollution and degradation but not always in the ways that environmentalists believe. Environmental rhetoric suggests that man has ultimate control over the creation as if the human race could bring the world to an end by its own power. Scripture asserts that God, not man, is sovereign. The Bible does teach that man is responsible for environmental turmoil. If men keep covenant with God, living in obedience to Him, God promises to bless them and their world with productivity, safety and beauty. (Deuteronomy 28:1-14)

The creation itself is "on God's side." Now since the sanctuary is not only the center of the world, but also a model of the world, we can form a Biblical idea of environmental pollution from looking at sanctuary pollution. We can see that sanctuary pollution has three effects: It makes the sanctuary unsafe, unfruitful and ugly. We can define environmental pollution in similar ways.

Pollution is a foreign substance that causes a resource or a part of the world to become dangerous, unfruitful or ugly. For instance, polluted water is water that is unsafe to drink or water that is incapable of supporting plant or animal life. Polluted land is land that is unsafe for direct contact, perhaps from chemical dumping or overuse of chemical fertilizers; it may be dangerous and ugly from litter or overuse. In the same way, polluted air is unsafe to breathe. A polluted environment is one that is unsafe to live in, unproductive and ugly.

An example of this is the small town of Mbita, where we have been ministering. All the streets are covered in trash. The town's beaches, where fishing boats come in to offload their catches, are filthy. The town trash dump, as well as the slaughterhouse, is on that beach. Seepage from both these sources enters the lake there. We have witnessed people defecating on the beach, as well as children swimming and urinating in the water. This is the same point where the local hotels come and collect water for drinking and cooking! No wonder the water-borne diseases in this area are so high! Bilharzias is rampant in the community, as well as typhoid and dysentery. It has also been shown that the rate of malaria is escalating because mosquitoes breed in the trash at the dump and along the roadsides. Seventy percent of mosquitoes in the area are bred in man-made containers.

Civic governments all over the world need to have a greater role in regulating pollution and to be concerned with laws that deal with pollution and promote environmental care, including the development of "green belts" under their jurisdiction. City and

> It is essential that we find a way to make humans more attentive to the earth as well as to all of the earth's residents, present and future. This is the aim of sustainable development, and the responsibility of every individual.
>
> — Christian Broodhag,
> *Earth from Above*

town planning is also part of the Great Commission. A productive, safe, and beautiful environment is a blessing of the Lord. Scientific study of environmental problems is important and should be part of a Christian's involvement, but in the end no amount of technical expertise will shield rebellious men from the judgment of a holy God. Humanity's twin imperatives given by God are multiplication of born-again believers and management of the earth. As we evangelize and disciple we need also to help people manage their resources. We need to preach a two-handed gospel. We need to help in restoring the sanctuary that God has given to us to steward.

Chapter 23: **A DYING LAKE**

*"Thus says the Lord God, 'On the day that I cleanse you from all
your iniquities, I will also enable you to dwell in the cities, and the ruins
shall be rebuilt. The desolate land shall be tilled instead of lying desolate
in the sight of all who pass by. So they will say, "This land that was
desolate has become like the garden of Eden; and the wasted, desolate,
and ruined cities are now fortified and inhabited." Then the nations
which are left all around you shall know that I, the Lord, have rebuilt
the ruined places and planted what was desolate. I, the Lord, have spoken it, and I will do it.'"*

— Ezekiel 36:33-36

As I look out from the veranda of our base staff house onto
the wide expanse of water and islands of Lake Victoria, I am
struck with two dominant thoughts: the beauty of the character and creativity of God and the foreboding reality of a lake dying
through humankind's misuse and ignorance.

The sights and sounds seem idyllic: The sun sending off brilliant
streaks of red behind Mfangano Island and into Uganda in the
west. The cry of the African fish eagle claiming its territory comes
across the evening breeze. Near our jetty I can see three large black
shapes in the water and hear the familiar "mmhaa-hu-hu-hu" of
our resident hippos. As the sun disappears, the quick transition of
an African equatorial day into night reveals lights of many fishing
canoes across the horizon, transforming what was a vacant expanse
of water into a city of lights.

An example of God's perfect plan in creation is displayed in this
diverse ecosystem of Lake Victoria. Here from the hippopotamus
to the haplochromines, from pelican to protozoa, food chains and
webs blend into a dynamic system. Even the sinister crocodile has
its place, maintaining balances. Certain haplochromines keep the
algae content down and others check the lake fly populations.
The convoluted shoreline contains innumerable shallow bays and
inlets, as do hundreds of islands within the lake. Many of these

geographical forms – swamp, papyrus beds and wetlands – differ appreciatively. Alas, many of the checks and balances in these diverse systems have disappeared at the hand of humans, and if people in this region are going to survive, the diverse systems will have to be restored.

In 1858, the British explorer John Speke discovered the southern shore of the lake, named it Victoria after his queen, and proclaimed that he had found the source of the Nile. In 1875, Henry Morton Stanley, a reporter for the New York Herald, was sent out to find David Livingstone. When he found the doctor, Stanley uttered the famous, "Dr. Livingstone, I presume." Stanley circumnavigated Lake Victoria to confirm Speke's claim. Stanley, a disciple of the deceased Livingstone, sent word back to England, asking for missionaries.

With the missionaries came traders and soldiers, and the region was soon colonized into what is now Uganda and Kenya. In the meantime, Germany colonized Tanganyika, now Tanzania, the third country that borders Lake Victoria. By 1902, the colonial government in Kenya had built a railroad from Mombasa on the Indian Ocean to Kisumu on the lake, and soon Europeans laid bare vast tracts of forest in Lake Victoria's watershed to plant coffee, tea, sugar and cotton. As the human population increased and tribal people such as the Luo and Subas gave up their nomadic ways, they increasingly turned to the lake not only for sustenance, but also to satisfy a market for fish in the growing urban centers and overseas.

THE COLLAPSE OF HER MAJESTY'S LAKE

Lake Victoria, a shallow lake with a maximum depth of 80 meters, 68,800 square kilometers of surface area, and 180,000 square kilometers of adjoining catchment, is the world's second-largest freshwater lake and the largest lake in the developing world. At one time it held more than 400 species of extraordinarily diverse endemic fish, the haplochromines, and inspired scientists around the world. Fishing in Lake Victoria intensified in 1905 when the

British introduced flax gill nets and overfishing caused catch sizes
to drop. Fishermen began to use smaller and smaller mesh sizes to
increase their catch. Obviously this drastically reduced the number
of breeding adults and juvenile fish of many native species. By the
1950s, British officials decided to restock the lake with new fish,
the first non-native fish being the Nile tilapia. In 1954, Nile perch
were introduced and were restocked in the 1960s, despite opposi-
tion of scientists who feared that the lack of natural predators
would ultimately destroy the system.

A survey sponsored by the United Nations and completed in
1971 showed that the haplochromines still made up 80 percent
of the lake's biomass. From 1974 to 1979, Ugandan dictator Idi
Amin cut off access to Lake Victoria so no research was done. In
1980, the Nile perch leaped to
80 percent of the fish biomass of
Kenyan waters, and the haplochromines to less than 1 percent. Ugandan and Tanzanian
waters also showed this devas-
tating change. The commercial
catch of Nile perch in 1978 was
still less than 5 percent, but by 1990 it jumped to approximately 60
percent Nile perch and 40 percent omena, a small local fish about
the size of a sardine. The haplochromines and other fish had virtually disappeared from the commercial catch. The theory is that
the haplochromines were simply eaten as food for the introduced
Nile perch. Because of the loss of other food, the larger Nile perch,
which reaches 6 feet in length, have turned to consuming smaller
specimens of its own species, threatening the very industry that it
was brought to aid. Consequently the main source of food for the
local human population is diminishing alarmingly.

Nutrient inputs to the lake are three times what they were in
the early days, most of the increase occurring since 1950. This
has caused a five-fold increase in algal growth, resulting in human

> A child is like a flower
> which will blossom and
> mature in conditions
> needed for its growth.
>
> — Don Richards, from
> *Reflections in Wilderness*

and animal illnesses, more water treatment for urban centers and clogging of water intakes. The decay of massive algal blooms at the lake bottom causes deoxygenation, resulting in a nearly total loss of deep-water species. Periodic upwelling of the hypoxic water has also caused huge fish kills of shallow-water species.

The eutrophication spurred the spread of the introduced water hyacinth, which appeared in 1990, floating down from Ugandan rivers. It grew at a rate of seven to ten acres per week, interfering with boat traffic and fishing. Because it grew so densely, one could actually walk on some areas of the lake, and in some parts no water was visible to the horizon. Advancement of the water hyacinth closed Kenya's main lake port, Kisumu, in January 1997. By August 1997, Homa Bay, some eighty miles south of Kisumu became infested and closed in. This infestation affected water for industrial and household use and also affected the oxygen level in breeding waters that led to the death of many fish. Work to eradicate the hyacinth provided temporary relief, but the infestation had returned to high levels by 2006.

> Desertification affects
> more than 1 billion people.
> — www.unep.org, 2006

Pollution is on the increase. Each night in Kenyan waters, more than 160,000 fishermen are out in some 40,000 fishing boats. Each boat carries four fishermen and four paraffin pressure lamps. These lamps are placed on small papyrus rafts connected by a line to which nets are attached, forming lit-up streets of nets to which lake fly, other insects and fish are attracted. Apart from the overfishing, the fisherman use the lake as a bathroom each night, causing e-coli contamination from the human feces. Industries such as breweries, abattoirs, tanning, and fish processing pollute the shoreline and rivers that feed the lake. Small-scale gold mining contaminates the lake with mercury. Thus the lake is used as a source of transport, energy, food, and drinking and irrigation water and as a cache for human, industrial and agricultural waste. The rate and magnitude of changes in the lake ecosystem have had alarming ecological repercussions and social consequences for more than 30 million people who depend on the lake. Scientists around the world agree that if significant and effective steps are not taken quickly, Lake Victoria will cease to sustain life. The death of this lake would bring unparalleled suffering to one of the world's fastest growing populations.

THE VICTORIA LAKE BASIN

The shoreline, valleys and hills that surround the lake were twenty years ago well-wooded savanna, including large trees such as Balanites aegyptica, Acacia seyal and Ficus sycamorus. The upland regions were covered with trees, shrubs and tall grass. The mountain reedbuck and southern reedbuck, duiker and bushbuck thrived, as well as monkeys, hyena, leopard, mongoose, porcupine, hedgehog and hares. The heads of millet grown by the local populations were so heavy that they broke their stalks with their weight and some of the crop was left standing in the field because more grew than was needed.

Peter Ollimo, a local Luo and a co-worker, describes the changes that took place around 1980. "The population dramatically increased due to larger family units and people moving from the islands to the mainland. Government schemes to settle the region and rid the area of tsetse fly included cutting down vast areas of forest and bush. The nomadic pastoral Luo people, a Nilotic race, became sedentary and still maintained herds that are too large for the land to sustain. Indiscriminate burning of the area, cutting trees for charcoal and over-grazing by cattle and goats have turned this once fertile region into a semi-desert."

Ignorance of appropriate farming techniques, including contouring and water table conservation, have amplified the semi-desert conditions and caused perennial streams to dry up. The water table has all but disappeared and bare hillsides erode, silting the riverbeds and the lake. Tall waving grass and excellent crops of millet and maize have given way to bare rock and a semi-arid landscape, not able to adequately support the people.

ENDANGERED SPECIES

In 1998, the World Conservation Union Red Data Book of
Endangered Species listed the hundreds of endemic fishes of Lake
Victoria under a single heading: "Endangered." Now the people of
that region are endangered!

Our staff at the YWAM Mbita base feel the devastation per-
sonally. Kennedy Omalla, our boat captain and a member of the
leadership team, lost his wife, Emily, in the much-anticipated birth
of their first child in 1995. Emily was diagnosed with cholera, typhoid, amoebic dysentery and malaria in her ninth month of preg-
nancy. Both wife and child died in the hospital. (Kennedy is now
happily remarried and is the proud father of three lovely children.)
Most hospitals and clinics, where they exist, are under-equipped
and understaffed and cannot cope with the demands. Mortuaries
are overflowing. Conditions in the region have led to an unprec-
edented death rate, mainly from water-borne diseases. This region
also has the most AIDS deaths in Kenya, a result of weakened
immune systems, the transient fishermen, prostitution, and the
cultural practice of polygamy, producing an alarming number of
orphans. HIV is as high as 42 percent and on some of the islands as
high as 70 percent. Malnutrition is rife and infant mortality immensely high.

In Genesis 1 and 2 we learn that God made a perfect world with
which He was very pleased. Now because of sin and transgression
against God's commandments, Lake Victoria, its environs and its
people are suffering. In Genesis 1, God says that we are created
in His image. God commands us in Genesis 2:15 to look after His
creation and to tend the garden. We who are part of His church
should work for the restoration of Lake Victoria and the part of creation dependent on that majestic body of water. The church worldwide needs to address not only evangelism and discipleship but also
the ministry of environmental stewardship on a global scale.

Time is running out on the lake like water draining from a
broken pipe. A World Conservation Union program is working on

the Lake Victoria problem. The three countries that share the lake, Kenya, Uganda and Tanzania have launched the cooperative Lake Victoria Environmental Management Program. These initiatives are leading in the right direction, but there is an urgent need for a combined effort at the grass roots level involving communities, churches and schools. A sustainable program is needed to restore the majesty and function of Africa's great lake and the dignity and well-being of its people who are in truth the pinnacle of God's creation.

As a result of what is going on in this region and through the Lord's direction, the YWAM base at Mbita, now in its 11th year, has been actively engaged with community development work through the local schools, churches and fishing villages. With the schools the base runs biblically based environmental education camps, encouraging the students in projects back in their communities that will benefit those communities. A number of the schools have established tree nurseries and replanting projects in the water-shed. Others have learnt how to make solar cookers that eliminate the need to cut down trees for fuel.

> Listen to the sermon preached to you by the flowers, the trees, the shrubs, the sky, and the whole world. Notice how they preach to you a sermon full of love, of praise of God, and how they invite you to glorify the sublimity of that sovereign Artist who has given them being.
>
> — St. Paul of the Cross, 1694-1775

The base also offers courses and schools in environmental stewardship, which includes instruction in alternative agricultural methods and in agro forestry. The courses include appropriate technology, health and hygiene, teachings on malaria, and clean water issues. Seminars are also offered to pastors to unify the body of Christ, and one staff member has established a Bible college on Mfangano Island.

In 1997, Nancy launched Christ Gift Academy, a school for
orphans and children from poorer families. Starting the first year
by using a local church as a facility, the school by 2006 had developed as an elementary school up to eighth grade with more than
300 children and a staff of 40. Land next to the base was donated
by community members, including Chief Elly Onundo. Under the
guidance of Steve and Judy Cochran, a whole campus has been
developed. The school has produced excellent results and has been
an example to all in the area of the potential of orphans and the
poor.

What started as a vision from God with humble beginnings in
1995, on three-and-a-half acres of land has now extended to a
major ministry on nearly 15 acres.

The vision has now extended to Maasailand with Andre Brink
empowering and enabling the Maasai. Plans are afoot through
Marine Reach to launch a YWAM mercy ship on Lake Victoria.
This Uzima project, under the leadership of Lyle Hall, has a vision
to use the Mbita base as the home port for the ship and to be able
to reach out to the whole Lake Victoria basin. His vision is also
beyond Lake Victoria to reach out to all the great lakes of Africa
with mercy ministries. A start has already been made on Lake Tanganyika at Kigoma, which was originally Ujiji, David Livingstone's
headquarters.

In the long run, YWAM Mbita is earnestly trying to train and
build up Christian leaders and caring communities that will look
upon their environment as a gift from God to be nurtured and
cherished. As they tend the lake and the land God has given them,
the prayer is that God will bless them and restore the area into a
garden that is abundant in provision and a beautiful reflection of
the original handiwork of God.

Chapter 24: **WORLDVIEW**

"For since the creation of the world His invisible attributes are clearly seen, being understood by the things that are made, even His eternal power and Godhead, so that they are without excuse."

— Romans 1:20

'We are without excuse because God has revealed himself to us through His creation.

"Every star is an announcement. Each leaf a reminder. The glaciers are megaphones, the seasons are chapters, the clouds are banners. Nature is a song of many parts but one theme and one verse: God is.

"Creation is God's first missionary. There are those who never held a Bible or heard a scripture. There are those who die before a translator puts God's Word in their tongue. There are millions who lived in ancient times before Christ or live in distant lands far from Christians. There are simple-minded who are incapable of understanding the gospel. What does the future hold for the person who never hears of God?

"Again Paul's answer is clear. The human heart can know God through the handiwork of nature.

"The problem is not that God hasn't spoken, but that we haven't listened. God says His anger is directed against anything and any one who suppresses the knowledge of truth. God loves His children and hates what destroys them."

— From *In the Grip of Grace* by Max Lucado

Although God's character and ways are clearly reflected through creation, our perceptions of these are often distorted by our worldview. Our view of God, ourselves, and the world around us cannot in reality be separated. If we are looking to impact societies across cultures, we must look at worldview in three major areas. We must know our personal worldview, which usually is derived from our nation's dominant culture in our own subculture. Secondly, we need to study the worldview of the people with whom we are living and working. Thirdly, we as Christians must become consciously Christian, repenting of our false views and embracing God's reality.

All people in the world have particular worldviews that accord mostly with the culture they were brought up in. How they view the world shapes their development, prosperity or poverty more than does their physical environment. Each worldview produces different values. Most worldviews are passed on from parents to children.

Worldviews have consequences. For example, whilst we were working around Lake Victoria in East Africa, we found that many people on remote islands had never been on the mainland. Their worldview was shaped by what surrounded them and the beliefs in their confined area. Beliefs and religion, cultural behaviors and thinking are passed on from generation to generation. This was their world and they grew up to believe that the outside world must be like theirs, that their culture and beliefs were universal.

I have noticed that people's worldviews affect their physical environment. For example, the forests around Lake Victoria have been cut over for firewood, leading to erosion of topsoil, a lower water table and a lack of clean drinking water. When questioned, most people believed it was their job to cut down trees and that God or whoever had created trees in the first place would create more trees. It was not their responsibility to plant trees! This is a result of an animistic worldview where everything in life is cyclic. Seasons come and go each year, there is life and death, and there is no real planning for the future. This fatalistic worldview also leads to hopelessness.

We can compare three worldviews as Darrow Miller describes them in his excellent and helpful book, *Discipling the Nations*.

(1) BIBLICAL AND CHRISTIAN WORLDVIEW

Christianity's worldview is consistent with reality through Biblical scripture. Paul told believers to bring every thought captive to Christ. (2 Cor. 10:5) Every area of human life should be under

Christ's lordship, and all culture should be redeemed for God's glory, the establishment of God's Kingdom on earth and the successful development of nations.

Theism is rooted in the ancient Near East. Theism sees ultimate reality as personal and relational. God exists. He created a universe of physical and spiritual dimensions. Truth, as revealed by God, is objective and can be known by man. God's character establishes absolute morals. There is only one personal-infinite God, and He created man and the world.

(2) SECULARISM OR SECULAR HUMANISM

This is a worldview that sees reality as ultimately physical. Having evolved to the highest level, man is in charge of the earth. Morals are relative, and there are no moral absolutes. Social consensus establishes values. The secular world and the spiritual world are separate. Jobs and church are divided into compartments. There is no room for religious or spiritual concepts in man's day to day work. Secularism has become materialism. Man is the center of the universe, and God might be out there somewhere! There is no real meaning or purpose to life.

> Deforestation can be the result of deliberate removal of trees for agriculture or urban development.
> — http://wikipedia.org

(3) ANIMISM

Animism views reality as absolutely spiritual. The physical world is an illusion, and the real world is not apparent. Spirits animate the world. There is oneness of all spirit. The New Age religion comes out of animism. It is derived from the Far East and world folk religions. Animists believe in millions of gods and ancestral spirits. Bad things happen when the gods are angry. This leads to fear and constant appeasement. Man's goal is to survive an endless cycle of existence and to escape the world. Reincarnation is a central belief. Hinduism and Buddhism fall into this category.

THE POOR AND THE ROOT OF POVERTY

Apart from catastrophic events such as war or natural disasters, tsunamis, earthquakes and hurricanes, poverty is not just a chance occurrence. In most instances, poverty is a result of how people look at themselves. They grow up with a poverty mentality. They see the world through a mindset of poverty.

These poverty values are caused by the lies of our enemy, Satan. The poor are often fatalistic. They reason that there is nothing they can do about their lot. They believe they are destined to be poor and always will be. There is no clear concept that they are created in God's image, created with a mind and a good pair of hands. Sometimes they say, "I am poor because others make me poor." They do not rise above these thoughts.

As we read the Bible, we can see quite clearly that being poor is not in itself a sin. God has a very special concern for the poor, and He often warns that wealth can become a danger. What we need to understand is that God did not desire poverty for man.

In Deuteronomy 15:4-5, Moses tells the people, *"However, there will be no poor among you, since the Lord will surely bless you in the land which the Lord your God is giving you as an inheritance to possess, if only you listen obediently to the voice of the Lord your God, to observe carefully all this commandment which I am commanding you today."*

God's ways bring prosperity. A mindset that is rooted in poverty often resists development.

CREATION: AN OPEN SYSTEM

Secularism and animism look at the world as a closed system — believing that there is not an infinite God that can intervene in the course of events. According to the secularists, man is in control, God is left out of the picture. Their belief, for instance, in the overpopulation problem now leads to abortion and the killing of millions of babies worldwide. From this perspective of a closed system, resources are limited and will run out with too large a population. It sees people as the problem and not the solution.

THE TRUTH ABOUT STEWARDSHIP

Secular stewardship	Christian stewardship
Earth was created by chance.	Earth was created by design.
Mankind evolved by chance.	Mankind was created in God's image.
Evolution gave man attributes that allow him to dominate.	God gave man dominion.
Only people can "care" for the earth.	God cares for His creation.
Only people have the power to influence the fate of the earth.	God is sovereign.
Survival is the responsibility of mankind.	"The earth is the Lord's" and man in His steward.
Earth is all there ever will be.	There will be a new Earth.
Love is a fleeting human emotion.	Love is an act of the will.
Earth is under the curse of mankind.	Earth is under the curse of sin.
There are no absolute standards.	There are absolute standards.
Death is the end of existence.	After death we will be judged.

Biblical theism puts God on the throne. He is not only infinite, He is also personal, and the universe that He created is both spiritual and material. Because He is a creative God, He has put creativity into man. The world is governed by God and His natural laws that are open for man to discover and utilize. Therefore it is a marvelous open system, where God can and does intervene for His pleasure and designs and in answers to prayer.

In all of this we can see that man's stewardship of which God speaks in the Bible is essential so that creation can become increasingly fruitful. As ambassadors of God on earth, we have been given the awesome task of bringing forth all the potential of creation. We are commanded to be wise stewards and caretakers of His garden — planet earth and everything that is on it. Not only that, but we can improve conditions by being open to God's creativity, having God's vision and power to make it happen.

Ezekiel 36:36 says, *"Then the nations that are left all around about you shall know that I, the Lord, have rebuilt the ruined places and planted what was desolate. I the Lord, have spoken it and I will do it."*

THE KINGDOM OF GOD — DOMAIN STEWARDSHIP AND ECONOMICS

When we re-create cultures, we are re-creating the Kingdom of God. *"Your Kingdom come, Your will be done on earth as it is in heaven."* (Luke 11:2b) The dominions of church, community, government, economics, arts and entertainment, science and technology, and education are all areas of human concern. As Christians we should be influencing all these dominions. We cannot afford to leave out any of them. Even if a person does not believe in God, that person is still made in His image. Revelation 11:15 says, *"The kingdoms of this world have become the kingdoms of our Lord and of His Christ, and He shall reign forever and ever!"*

Jehovah Jireh is our provider. If we do not understand that, we will believe in limited resources and over-population and buy into the humanistic basis of world economics. This is not Biblical. The Bible speaks of a God with unlimited resources. Solomon observed that he had never seen the righteous begging bread (Psalm 37:25), and the story of Elijah and the widow (1 Kings 17:8-14) illustrates God's provision.

There is no need for any nation or community to be undeveloped. When you change the assumptions of economics, you change communities. If people think poor, they remain so. The concept of limited resources develops greed, poverty, corruption, communism and fear. As Christians in communities, we need to deal with this misconception and provide a better answer.

The basic goal of biblical economics is enablement. A good example of this is small businesses and business loans (see Deuteronomy 15:1-11). From the beginning of the Bible to the very end of the Book, God wants to bless everybody. Why is this not happening more often?

We are responsible for our neighbor ("Love your neighbor as yourself"). We are meant to help others and need to come out of our selfish modes. It is not possible for each of us to help everyone, so we need to start with the family. The church's obligation is to feed the widow only if she has no family to look after her. Our obligation is to apply God's principles in every area of life. We bring God's kingdom in first with the individual, then the family, the community, our nation, and then other nations. It is a system of multiplication and expansion.

"But you shall receive power when the Holy Spirit has come upon you; and you shall be witnesses to Me in Jerusalem, and all Judea and Samaria, and to the end of the earth." (Acts 1-8)

We are to help the poor amongst us, and we need to start with something that will make a difference. Giving money is not the answer, but helping people to make their own living is. This principle is one that we have attempted to apply with our ministry in Kenya. Jesus says that we will always have the poor amongst us, and these are often poor through poor choices. Every person is made in the image of God. Therefore unless handicapped from birth, each person has God's creativity. Each person has a brain and hands and feet and the ability to improve his situation.

It is not wrong to make money, but the way we do it affects our nation. Greed is guiding the money making in many parts of the world. If a nation cannot get outside aid, it has to develop its own

resources, using its own citizens' giftedness. When a nation bor-
rows from other countries it loses motivation and confidence, and
its citizens become aid dependent. The Bible speaks of individuals
being able to borrow but says that nations are not to (Deuteronomy
15:6). God wants us to know that we can create with Him. Dependence leads to a lack of initiative. So there should be no free lunch.
Our role as missionaries and the church is to help the poor to help
themselves. This can be done, for example, by teaching, facilitat-
ing and setting up simple repayment schemes. This whole process is
part of enablement and discipleship.

> Using Biblical principles,
> prosperity increases as
> individuals are able to
> release their true, God-
> given potential.
>
> — Loren Cunningham,
> *The Book that Transforms*
> *Nations*

The right to own is a biblical principle. God was teaching
Israel that each should own
land. He says in Deuteronomy
72 times, "Take the land!" We
need to encourage land ownership and wise stewardship.

Economics is tied up with
good planning. How are you
going to market? What are the
state of your roads and communications? A number of the people
in Mbita, Kenya, have started their own wayside kiosks, most of
them selling vegetables and fruit imported from Kisi, almost 100
kilometers away. The cost of the produce is elevated by transportation on roads that are so bad that vehicle upkeep becomes a prime
concern. Often in wet weather the roads to Mbita are impassable,
and it is impossible to ship goods in or out. The church in Mbita
can make it a better place to live by getting involved in this whole
process.

People need to make a profit. If they eat everything they produce, they go into debt. Then they will never be able to get a
good education nor develop beyond a subsistence level. A person
without a concept of ownership cannot be a wise steward. Israel
was taught not to destroy the material blessings of the land but to

conserve what was there. God gave instructions on what to eat and what not to eat. (See Deuteronomy chapters 12, 14 and 15)

Deuteronomy chapter 23 is also full of hygiene laws. God cares, but do we? In the U.S.A. today, almost 50 percent of solid waste goes back into the water supply. In the Smoky Mountains where we now live, there are huge amounts of trash thrown along the roadsides. This is a tourist destination and people want to visit the area because of its natural beauty. However, tourists are becoming disillusioned because of the amount of trash. If they stop coming, it would have a huge impact on the economics of the area, as almost every business revolves around tourism. I believe this should be a concern of the churches in the area. We can get so heavenly minded that we cannot see the good earth that God gave us to look after.

We cannot change people, unless they know God and see their worth in His eyes. As they get to know God, their attitudes will change toward their environment, and they will become part of the solution instead of part of the problem. They will become involved from a grass-roots level. Changes from the top often will not last, but if these changes rise out of the community, they will last. We as Christians must live out the principles of all domains. We need to be concerned.

Chapter 25: **BAPTIZED IN LAKE VICTORIA**

*"He who believes and is baptized will be saved; but he who does not
believe will be condemned."*

— Mark 16:16

I first met Kennedy Omala in February 1995 when Andre and
I were spying out the land and lake near Mbita in the Nyanza
Province of Kenya.

Naphtali Mattah, director of the Suba Bible Translation and
Literacy organization (BTL), had invited us to cross the lake with
him to Mfangano Island where he was stationed. Naphtali, a native of Mfangano and a Suba, was in the process of redeeming the
identity of the Suba tribe and their language. The Subas had been
overcome by the Luo generations before. One of his main goals was
to translate the Bible into Suba.

Landing on Mfangano, Naphtali led us to his office, a small mud
hut with a solar panel on the roof that enabled him and his small
staff to run a computer! We were amazed to see this technology
working in such primitive conditions.

Kennedy Omala happened to be the pilot-driver of the BTL boat
that took us across the lake. Kennedy knew Lake Victoria like the
back of his hand. In 1997, he joined Kenya Islands Mission and
later became the boat captain of our 42-foot-long wooden boat,
"Ebenezer." When Ken tragically lost his wife during childbirth,
Andre took him into our home. During that time he related the
following experience that had led him to Christ:

"During the year 1994, I was co-pilot on a BTL boat during a
night voyage from Homa Bay on the mainland to Mfangano Island.
I began to realize that the pilot did not quite know the direction.
After about one-and-a-half hours on the water, I stood up to check
the course. As I did so, I saw something coming straight for us out of
the darkness. I shouted a warning, and our pilot tried to maneuver
our boat out of the way. Unfortunately, the oncoming boat corrected

in the same direction and hit our stern, causing us nearly to capsize.

"As the thirty-six passengers on our boat panicked, they all shifted to the same side of the boat, and I was knocked overboard. My immediate thought was to grab the engine, but it was still running, and in fear of being cut by the propeller, I let go. The weight of my clothing was too much, and I realized that I was sinking. I struggled successfully to remove some of my clothes, and I rose to the surface.

"I was totally confused and looked around for some help. I realized that the pilot and passengers on my boat had not noticed that I was missing. It was then that I spotted the other boat and called for help from them. I later found out that the co-driver of the other boat had heard me calling, but he was so scared that he had killed someone that he decided to make a run for it.

> The hour is late, the need is urgent; God is stirring His people to break down dividing walls so that we might work together in unity to "tend his garden."
>
> — Tri Robinson, *Saving God's Green Earth*

"By this time my boat was about thirty meters from me and still moving away. The pilot then stopped the engine, thinking that there might be damage to it. It was then that they heard me calling for help and shouted back that they were coming back for me. It was dark, and the weight of my shoes and remaining clothes started pulling me down again, so that my boat passed me twice without spotting me. It was during this time that I clearly heard the Lord say, 'Take it easy, don't panic, I am the Lord.' I came back to the surface and shouted to the pilot to stop the boat and wait for me to swim to it. I was pulled aboard and then continued to Mfangano Island, which is my home."

After this incident, Ken became the pilot of the BTL boat and subsequently gave his life to the Lord. God had prepared his heart to serve Him during that ordeal. It seemed almost like baptism before commitment. We are aware of how the Lord has wondrous ways of bringing His children to Him!

Chapter 26: WALKING WITH MAASAI

I got to know Andre Brink during the EARS school in the Amazon in 1994. He was a fellow South African, and we shared a common interest in things wild and in the redemption of God's creation, including mankind. Andre's pioneer and creative spirit shone through as a student in the EARS school, although he was only 19. While listening to God's voice, he, like I, had a call to minister to the indigenous people of Kenya in the area of discipleship and environmental stewardship.

Through a series of divine appointments, God led both of us to spy out the land in Kenya in February 1995. This resulted in pioneering Kenya Islands Mission (KIM) on the shores of Lake Victoria. During that same year, the two of us camped for the first time in the Maasai Mara, an extension of the great Serengeti plains. Here the Maasai still live as they have for hundreds of years, together with their cattle and the wild animals that inhabit the area. As we made more trips into the Maasai Mara and Maasailand in the years that followed, God sowed a seed in Andre's heart to work with these warrior people and to set up a base camp in Maasailand concentrating on discipleship, environmental stewardship and education.

This opportunity came in 2000, the same year Nancy and I had another divine appointment at the Mennonite Guest House in Nairobi. Seated at the dinner table with us were a group of Christian Missionary Fellowship (CMF) missionaries who were ministering in Maasailand. During the course of conversation, Nancy and I brought up Andre's vision of working in Maasailand and the need for a site where this could be realized. It turned out that CMF was moving a number of missionaries out of Maasailand and that a few of its mission stations would be vacant. After a number of meetings with their committee, the mission station at Tiamanang'ien in the Loita Hills, adjacent to the Maasai Mara and close to the Tanzanian border, was given to Andre' to begin the fulfillment of God's vision.

The Loita Hills of Kenya are one of the last true unprotected wilderness areas in Maasailand. This diverse and mysterious land-scape is a place where Africa still holds true to its wild character. At an elevation of 7,000 feet, the hills and valleys are covered with a mixture of Afro-montane forest, glades, wetlands and acacia thorn veld. The Naimina Enkiyio Forest (Forest of the Lost Child) supports a rich diversity not only of plants but also birds and numerous species of animals. Here animals such as elephant, buffalo, hippo, lion, leopard, and antelope as well as baboon and colobus monkey share the forest and glades with a number of Maasai communities.

In ensuing years, Andre sent a number of Maasai to complete a DTS as well as an EARS school with us at Lake Victoria. As a result, one of them, Francis Yenko, has run more than nine DTS courses for the Maasai at the mission station at Tiamanang'ien; another, Parit Kashu, runs biblically based environmental stewardship courses with local schools; and David Koyie, a teacher himself, has embarked on a literacy project for Maasai warriors. Andre and these three wonderful Christian men now form the management leadership team overseeing these ministries.

> I wondered: How was it that man in all his seeking and striving should cut himself off from the creation? The birds, the animals, the mountains, the streams, the trees and the wind send out their messages for all to hear. Oh that we would have ears to hear!
>
> — Don Richards, *Reflections in Wilderness*

In 2004, God gave Andre and his co-workers a vision to set up a project called "Walking With Maasai," an eco-tourism ministry that would include a safari camp and wilderness trails in the Loita Hills, as well as camping safaris into the Maasai Mara. They reasoned that this project will generate income for the Maasai Discipleship Training School and to help fund the local schools and clinic, as well as generating capital for operation and growth of the overall project.

With this in mind, Andre and Kashu spent a seven-month internship with the Wilderness Leadership School in South Africa in 2004. Here they received further training in leading trails in big game country and have now been certified as bona fide wilderness and safari guides.

In 2005, one of the local communities gave the project 25 acres of beautiful wild country to establish the safari camp and use of the vast wilderness area surrounding it. Recent conservation strategy in the world has proved that conservation areas will be successful only if local communities benefit from it. This project should ensure many benefits to the local Maasai community, including employment and benefit to their livestock.

This vision grew out of Andre's love for Africa and a passion to restore and redeem to God what is His. The Maasai people of East Africa have always been the tribe that symbolizes Africa. Karen Blixen wrote in Out of Africa, referring to the Maasai: "Their style is not an assumed manner, nor an imitation of a foreign perfection, it has grown from inside, and is an expression of the race and its history." People from all over the world come to visit the great open plains of Maasailand and to get a glimpse of what Africa looked like when it was still new to the Western world. Today the people of the plains still roam the Great Rift Valley, grazing their cattle along with "God's cattle," as the Maasai call the more than 1 million wildebeest that share this vast land with them. They look and live much like they did 200 years ago, though much has changed over the last few years resulting in great damage to their culture, their land, and above all, their redemption. It is within this context that this vision, "Walking With Maasai," was born. *For it is God who desires all men to be saved and come to the knowledge of the Truth."* (1 Timothy 2:4)

This project already is acting as a community development effort to lift the people out of poverty and educate them at the same time. Another of Andre's desires is to redeem the culture and to expose tourists to truth as they go on safari with Christian guides. "Walking With Maasai" gives Christians the adventure of a lifetime out in the wilderness, while the fees they pay go toward the ministry and community projects.

As agricultural projects increase in rural Africa, it seems that most native people would much rather see the wild animals killed, because they compete for the land. One cannot blame this attitude since, in most cases, wildlife represents a nuisance factor for people, whereas agriculture even on a subsistence level, offers them survival and some sort of income.

This survival mode can be changed and wilderness areas maintained if the local communities benefit from the wilderness and the wildlife. Eco-tourism and wilderness trails in such areas where people live among the wild animals, offers ongoing income to the local people. In that way, preservation of wild areas and animals is seen as positive. This is what Andre and his co-workers are attempting to do. If this proves successful, poaching of wild animals would be minimized, and hopefully even eradicated, and wilderness areas protected.

With all this in view, what exactly is a "wilderness area"? There are a number of descriptions of it. From my point of view, a wilderness area can be defined as an extensive wild area where the local people can live in harmony with the land. A wilderness area maintains its wildness and should have minimal impact of modern civilization such as roads, vehicles and buildings. Apart from the local people who have already lived there, all tourist camps and facilities should be placed on the fringes and not in the wilderness area. Thus the area must be suited to a primitive type of outdoor recreation and should provide a feeling of isolation from the outside world. It should be a place where people can experience the wild as it used to be and where those entering should experience spiritual and physical recreation and inspiration. It should be a place where mankind can feel the presence of God the Creator, who made it all. There they can reflect on the Him as the source of their being and feel His presence and the peace that inhabits solitude.

DAVID KOYIE'S STORY

Whilst David was still in primary school in the Loita Hills, his father, Lempira Koyie, decided to move across the border into Tanzania. This decision was occasioned by the death of most of their cows, and Lempira believed he could acquire many more cows if he lived in Tanzania. Sadly, David went with his father, became circumcised into warrior life and together with his age-mates formed a Moramism, technically a band of raiders to steal goats and cattle.

> Hans Nielsen Hauge (1771-1824), preached that all we own and all that we are come from God. The Lord expects us to be good stewards of what He has given us.
>
> — Loren Cunningham,
> *The Book that Transforms Nations*

During their Olpul, a meat camp in the forest that traditionally lasted for three months, they decided to cross the Serengeti National Park at night to raid cattle on the other side. In the early hours of the morning they reached a community with many cows. David was asked to guard the gate of the cattle boma, whilst the rest drove the cows out. This meant that David would be the last to leave the scene of the crime, as his duty was also to shoot anyone with his bow and arrows who gave chase.

A chase it became, and it continued throughout the day and the following night. At first light on the second day a serious battle ensued, arrows and bullets flying everywhere. As they were fighting, David heard an inner voice saying, "David, why are you here?" Try as he might to ignore it, the voice kept repeating itself. The fighting continued now on an open plain until it seemed out of nowhere a ranger's helicopter and a National Park Land Rover drove into sight, both loaded with armed rangers.

As the helicopter and Land Rover closed in, seven of David's age-mates were killed. David suddenly felt guilty and alone. He continued running and soon found a hole in the ground that led to a tunnel. Heedless of the possible danger of snakes and other creatures, he crawled deeper and deeper in until it was pitch dark. A hand grenade exploded near the entrance. He expected rangers to follow him in, but all became silent.

Tired and helpless, he suddenly saw a bright light and once again heard the voice of God. "Open your eyes, David. There is a way out." The picture changed, and he saw a glittering cross. Closing his eyes he then saw a classroom with people of many races, all with Bibles. There was also one prominent red Bible. He saw himself being taught and then teaching others. He asked God. "How can I do this?" The Lord replied. "I will change you and teach you to reach out to the Maasai."

Opening his eyes, he looked to one side and saw another tunnel that showed daylight at the end! As he crawled from the tunnel, he realized that all was quiet. The authorities had recovered the cows. The realization came to him that God had saved him. On the spot he made a commitment to the Lord to serve Him.

Reaching his father's home, David decided to leave and return to his old school in Kenya. There the headmaster took him in. He passed the KSP exam as the best student in the school. As a result of this, the Ukeni Integral Development Project sponsored him fully to a teachers training college at Kericho, where he graduated to become a primary school teacher.

When Andre took over the mission base at Tiamanang'ien, David shared his experiences with him. Andre sent David to our YWAM base on Lake Victoria to complete a Discipleship Training School. On reaching the base with his wife, he was amazed to see that the classroom was the one that God had shown him in the tunnel; even the red Bible was there!

David is now teaching nineteen Maasai warriors in a literacy program, has started a church in his area and keeps getting more visions from God of how his project can expand. He comments to everyone that he meets that the Lord speaks to you, even as a nonbeliever, and that it is up to each person to be obedient and to carry out God's instructions.

THE TESTIMONY OF FRANCIS YENKO

The following is a testimony by Francis Yenko, another of Andre's Maasai co-workers:

"I grew up in a non-Christian family. I was the first-born and as such was entitled to own some goats and sheep, which my parents helped me to obtain. I loved my family, and I know that they loved me, but I depended so much on them, that God meant very little to me.

"My father used to drink a lot and then started to abuse my mother, as well as all of us children, whom he beat mercilessly. I started questioning the trust I had put in my dad. I became confused, but out of my confusion, my conscience led

> The initial step for a soul to come to knowledge of God is contemplation of nature.
>
> — Irenaeus of Lyons, 125-202 AD

me to think of God. A voice within me told me that God was real, He could help and that He was bigger than my dad. This was back in 1984 while I was in grade four, but unfortunately I did not take much notice of that inner voice.

"In school we used to have a class called Christian Religious Education (CRE), which eventually shaped my way of thinking towards God. Nevertheless, at the time I was still empty inside as I only had a shallow knowledge about God. Consequently, I became totally lost in those areas of our culture that are so contradictory to God's will. My life became a mess. I was out of God's will. I stole by force, I fornicated, and I became coarse and a slanderer, selfish, unloving and unforgiving. I became contemptuous of my parents. Through all of this, my conscience kept pricking me and pointing towards God as the only way.

"After I completed primary school, I started to see people going to church, and because there seemed to be so many of them and because of my conscience, I decided to go to church. I enjoyed the singing and also the praying, but somehow because of the many friends I had who were equally lost, I buckled under peer pressure. I still attended church but was not a believer. In 1988, a church was started in my home area by Christian Missionary Fellowship (CMF). Church was a gathering under a big tree; there was no building. As it was so close to my village, I started attending the services. All the teachings were by missionaries, and the teachings were incredible. The more I went and heard, the more responsive I became. Still, I had friends who were swaying me in the wrong direction, and my dad repeatedly forced me to herd the goats on Sundays. Missing church so much made me feel terrible.

"One Sunday in 1990, Dan, one of the missionaries, shared with us about how much God loves us and what a good friend we have in Him. This touched me so much that I went forward and gave my life to Jesus. I opened the door for Him, and I encountered Him in an overflow of happiness and excitement. I then attended baptism classes and was baptized. Strangely enough, as this happened, my father liked what he saw in me and told me not to worry about the goats and encouraged me to go to church each Sunday. Through the church, I received a New Testament Bible in Maasai, which I read avidly.

"From then on my life changed. God turned me upside down. As I repented of my sins, I chose to forgive others who had hurt me in some way. I became passionate to become more like Jesus by passing on His love to my family and neighbors. I asked God to forgive me of having been so self-absorbed, self-promoting, savage, addicted to lust and allergic to Him. I asked Him to help me reconcile with all people. And you know, God came through. He never let me down!

"Even though I continued to attend Christian seminars at Ewasuaso-ngiro, the CMF teaching center in Maasailand, I was not fully living out those teachings. God, out of His goodness and plans, brought Andre Brink from South Africa to Maasailand. I met him at Ewasuaso-ngiro, and it was he who introduced me to Youth With A Mission through which I eventually attended a Discipleship Training School at the YWAM base at Mbita on the shores of Lake Victoria. I went, thinking that it would be another course, just to attain knowledge, but what a mistake. Yes, I attained knowledge of God, but more than that, I learned to have a personal relationship with Him. Here He opened up my eyes and my heart to become His representative, servant and instrument in Maasailand and to see the need of discipleship in our local churches.

"Currently, I am looking forward to our ninth DTS at Tia-manang.ien in the Loita Hills as leader of the schools and as one of the teachers. God has blessed me with a lovely wife, Nolkorit, and with our first-born, a daughter named Yiamat. God is so good!"

David's and Francis' stories illustrate the principles of repentance, obedience and using the gifts that God has given us. Both of these men have received talents from God, and each is using them to fulfill the Great Commission.

FROM POACHER'S SON TO CONSERVATIONIST AND WILDERNESS GUIDE — KASHU PARIT'S STORY

"I was born and grew up in a very poor family. My father had four cows and ten goats, and that was all we owned. So for him to meet the needs of the whole family, he decided to change his occupation of being a pastoralist to poaching. His decision really affected our social structure as a family and more so my life as a boy. I never had enough time to get to know my father and to be with him for he was always away for his poaching expeditions in search of elephant ivory and rhino horn. My father's last poaching expedition was one to the Serengeti in search of a big tusker named Tembu. The safari that took him away from home for a month brought him back with severe malaria that claimed his life a week later. My father's death affected the whole family. Hopelessness reigned, for we depended on him for everything.

"I spent my childhood as a shepherd boy tending our small herd of cows in the field. As a Maasai child who was still living in wild surroundings, I had no interest in being sent to school. I loved the wilds and had my amount of time to learn from the book of nature what urban children must attempt to learn from poorly printed books! I witnessed nature's law of survival and its beauty, and through these unending lessons of the rich world around me, my interest and inspiration grew toward nature.

"In 1968, my mother suggested that I should go to school, something I was not keen to do. I never wanted to give up the freedom of the plains whilst looking after our cows. I also never liked what my friends, who went to school, told me about their encounters of school, being kept for hours in a closed room or being whipped by teachers. These were the things that never gave me an interest in school. It took my mother several weeks to convince me to go to school.

> *Still, still with Thee, when purple morning breaketh, When the bird waketh and the shadows flee Fairer than morning, lovelier than the daylight Dawns the sweet consciousness I am with Thee.*
>
> — Harriet Beecher Stowe (1853)

"My experience of school was boring, and I never enjoyed it until I was in grade six. By then I had good friends to play with and some captivating topics like the sciences that reflected the outside world which I was familiar with. With my searching heart of how all things happened to exist, I relied on science, which failed to explain the origin of life beyond the evolutionary theory.

"In 1997, just after completing high school, there was a seminar at one of the local churches that my mother had been attending for years. I decided to go and have a look at what was going on. After hearing hours of preaching, God touched my heart through one of the worship songs, and that day I gave my life to Jesus. The origin and purpose of life became so clear to me, and the natural world became a clear signpost that pointed to God, the designer and the creator of everything, all of which exist for His glory.

"When Andre Brink came to live and work in the Loita Hills, God led me to work with him in the area of environmental stewardship. He organized for me to do my DTS with YWAM at Mbita, and I then followed that up by completing the EARS school at the same venue.

"Through contacts Don Richards had with the Wilderness School in South Africa, Andre and I worked a seven-month internship with them and graduated as certified wilderness guides. At the moment, I am one of the directors of 'Walking With Maasai,' and I thank the Lord for the wonderful opportunities He has given me to rise up from a poacher's son to a conservationist and wilderness guide for Christ."

These last three stories of Maasai warriors becoming warriors for Christ are not only encouraging but also show that everything is possible with God.

Chapter 27: **STEWARDSHIP**

"What is this I hear about you? Give an account of your stewardship."
— Luke 16:2

A s the years have gone by and I have studied more and more of God's Word, He has been showing me how important stewardship is in His eyes. In fact God has appointed us stewards over many areas. These include:

■ The responsible use of time (1 Corinthians 10:31)
■ The responsible use of money (1 Chronicles 29:3, 5, 9)
■ Presenting our body to God (Romans 12:1)
■ Serving others with our spiritual gifts (1 Peter 4:10)
■ Using our homes for God (John 19:27)
■ Respecting and looking after his Creation (Genesis 2:15)

In Acts 17:25 it says, *"Nor is He worshiped with men's hands as though He needed anything, since He gives to all life, breath, and all things."* In fact, God is the owner, and we are His stewards.

Jesus emphasized the necessity of stewardship and the faithfulness required of God's servants in His parable of the talents. (Matthew 25: 14-29) Then Luke records, *"And the Lord said, 'Who then is that faithful and wise steward, whom his master will make ruler over his household, to give them their portion of food in due season? Blessed is that servant whom his master will find so doing when he comes. Truly, I say unto you that he will make him ruler over all that he has.'"* (Luke 12; 42-44)

But here comes the wakeup call:

"But if that servant says in his heart, 'My master is delaying his coming,' and begins to beat the male and female servants, and to eat and drink and be drunk, the master of that servant will come on a day when he is not looking for him, and at an hour when he is not aware, and will cut him in two and appoint him his portion with the unbelievers. ... For everyone to whom much is given, from him much will be required; and to whom much has been committed, of him they will ask the more." (Luke 12:45-46,48)

In Matthew, Jesus emphasizes that the smallest task in God's work may receive a great reward if we are faithful in performing it.

"For to everyone who has, more will be given, and he will have an abundance; but from him who does not have, even what he has will be taken away." (Mathew 25:29)

What this really illustrates is that a person must care for and use what God has given or else lose it. He is not just to bury it so that it serves no purpose, as the lazy servant did in the parable of the talents. This applies to abilities and spiritual gifts, as well as material possessions.

In 1 Peter 4:10, it says, *"As each one has received a gift, minister it to one another, as good stewards of the manifold grace of God."* Each talent that God gives us, He expects us to use to His glory and to help further His kingdom here on earth. Both David's and Francis' stories show how this principle works.

> The first thing we must understand is that environmental stewardship views nature as a resource and a provision.... God has given us His creation not to abuse but to use.
>
> — Tri Robinson, *Saving God's Green Earth*

If we have given our lives to the Lord Jesus Christ, then as His ambassadors on earth, we need to do all things as unto the Lord, no matter how big or small they are.

This is emphasized in Luke: *"He who is faithful in what is least is faithful also in much; and he who is unjust in what is least is unjust also in much. Therefore, if you have not been faithful in the unrighteous mammon, who will commit to your trust the true riches (spiritual matters)? And if you have not been faithful in what is another man's, who will give you what is your own?"* (Luke 16:10-12)

This applies to all that God has given us and particularly His creation. I mentioned before that the study of ecology is the study of our home or household. (Luke 12:42) This home is our planet Earth, created by God and of which He was very pleased. The word

ecology is derived from the Greek word oikos, meaning a home.
Ecology is thus the study of our home, planet Earth. The word
oikonomos primarily denoted the manager of a household, one
who looks after, and the word oikonomia means stewardship.

Have we as God's people, His church, been involved in the care
and stewardship of His creation? I believe that the church has not
given it much attention but has abdicated responsibility in favor
of the New Age movement. But it is our responsibility, according
to God's command to Adam in Genesis 2:15, to "tend the garden,"
to look after and use wisely this earth He has given us. We are to
redeem it to the glory of God who is the One who made it. This is
our challenge.

Chapter 28: **MAKING A DIFFERENCE**

T he challenge for many people is how they can make a difference. I believe that each one of us on this earth that God created has a responsibility to be a shareholder in restoring the creation to what God meant it to be, whether our contribution is big or small. Everything we do to protect God's creation is an act of love to Him whose eye is on each sparrow.

VALUES

Are your values in line with society's values even when they are detrimental to the environment? If so, you need to reassess and base your values on trust in God, a trust that incorporates action and cares for creation and its peoples.

USE OF RESOURCES

Are you using natural resources wisely? Whether it is in your attitudes; the use of God-given resources such as water, energy, and air; the protection of living and nonliving resources; or the handling of your finances, you need to be a wise steward of the little things that make a difference. Are you recycling?

HOW TO BECOME INVOLVED

First of all, go outdoors. Get your friends, family, business associates and church members out enjoying God's creation. This will inspire you to care.

Do not be wary of being involved with environmental movements, even if they are non-Christian. You can make wise involvement as a Christian a tool to witness to those who are not believers. Get involved in local and community affairs regarding environmental issues. If you are a teacher, you could start an environmental club in your school. Become part of a community service program that helps other people. They are the pinnacle of God's creation.

ENLIST ENERGY AUDITORS

Interfaith Power and Light will come to your church to assess its energy use and calculate the ways your church can save money on energy. You can use these savings to advance the kingdom of God.

REDUCE WASTE AND POLLUTION

In the U.S.A., we live in a consumer society, and in our consuming we tend to neglect the environment. It is estimated that the average American throws away seven-and-a-half pounds of waste every day. Yet some forms of reducing garbage and recycling are easy and can greatly change the impact we make on the environment. A study conducted by the Technical University of Denmark found that in 83 percent of cases, recycling is the most efficient method to dispose of household waste.

■ Don't use plastic or Styrofoam plates and cups. They are not easily recycled. The polystyrene foam takes a very long time to decompose in the environment and has been documented to cause birds and marine animals to starve. The California Waste Management Board Report finds that "in the categories of energy consumption, green house gas effect and total environmental effect, it is the second highest in causing problems." The EPA claims that "Chronic (long-term) exposure to styrene in humans results in effects on the central nervous system (CNS), such as headache, fatigue, CNS dysfunction, hearing loss, and peripheral neuropathy."

■ Many stores provide paper bags for your groceries. You can save them and bring them back to use several times. Or you can bring your own containers when you shop. Many stores recycle their plastic bags if you will take them back.

■ Don't litter. It causes problems. It is unsightly, can be dangerous to wildlife and can cause toxic damage. It is an irresponsible, selfish act.

■ Recycle aluminum cans. Every month, Americans landfill enough aluminum to rebuild the nation's entire commercial air fleet. Manufacturing with recycled aluminum uses 95 percent less energy.

■ Recycle your newspapers. Paper comes from trees. Americans use up nearly a million trees every week just through newspapers. Recycling a stack of newspapers just three feet high saves one tree.

■ Recycle paper in all forms. Americans throw away enough office paper each year to build a 12-foot wall of paper from New York to Seattle. Making paper from recycled paper reduces contributions to air pollution by 95 percent.

■ Stop unsolicited mailing. In the U.S. annually junk mail requires up to 18 million trees, uses 28 billion gallons of water to be produced, and would cost around $450 million to be taken to a recycling center. www.eco-cycle.org explains ways to stop this kind of mail.

■ Get involved in reforestation programs. According to UNEP, "The loss of natural forests around the world contributes more to global emissions each year than the transport sector. Curbing deforestation is a highly cost-effective way to reduce emissions." They are running the "Plant for the Planet: Billion Tree Campaign," which encourages the planting of trees in four key areas: (a) degraded natural forests and wilderness areas; (b) farms and rural landscapes; (c) sustainably managed plantations; and (d) urban environments.

■ Glass can be recycled indefinitely. It will never wear out. If glass is made from recycled materials, water pollution related to glass manufacture is reduced by 50 percent.

■ Recycle PET and HDPE plastic bottles. In 2005, 3.3 billion pounds of plastics that would have been thrown into landfills were recycled, and 52,000 workers were employed by the plastic recycling industry.

■ Most towns and cities in the U.S.A. have recycling bins marked appropriately. Marked containers for aluminum cans, glass, paper and cans may be kept at home or in your organization, including your church.

■ You can recycle cell phones to fund your ministries!

■ Buy recycled products, thus increasing the demand for them. In order for recycling to be economically viable, there must be a steady supply of recylable materials and constant demand for the reprocessed goods.

■ Make your own compost by recycling your food waste, such as vegetables, fruit, eggshells and potato peels. Do not dump any meat products into your compost. You can use your compost for gardening by making a compost heap. For further information in composting your kitchen waste and even on compost toilets, look up Jack Dody's website: www.christianhomesteaders.org

■ In your gardens, plant with diversity, a protection for plant and animal life. Use natural pesticides.

■ Buy organic as much as you can afford. This will increase the demand for organic foods, eventually lowering their cost and also protecting our land and water. Buying from local farmers has the same impact. Whenever possible, buy free trade items. Besides helping small holders, free trade products come from more environmentally friendly farms.

■ Save on electricity. Turn off what is not being used. Use energy-smart electric bulbs, available in most supermarkets, or become a solar user.

One weekday evening, in Bangkok, Thailand, city officials organized to have an electric meter indicating the city's use of electricity shown on all the major TV stations at 9 p.m. While the dial was showing, everyone was asked to go through their homes turning off all unnecessary electric lights and appliances not in use. The viewers watched the meter drop as electricity use was reduced by 735 megawatts, enough to shut down two medium-sized coal powered electricity plants.

■ For flashlights and clocks, use rechargeable batteries and buy a simple recharging unit. This will protect the environment from the toxic mercury used in batteries.

■ Discarded items: Americans are noted hoarders and sometimes we dump what we don't want, such as old clothing, furniture, old garden tools, kitchen appliances, etc. Rather, take them to a neighborhood thrift store or the Salvation Army.

■ Water: Less than 1 percent of the world's total water supply is potable. When you brush your teeth, do not keep the water running. When you shower, turn off the faucet whilst you soap your-self. The Global Tommorrow Coaltion has created a helpful comparison chart on how to save water:

Saving water	Wasting water
SHOWER	
Wet, soap, rinse: 4 gallons	Regular shower: 25 gallons
BATHING	
Don't use tub	Full tub: 36 gallons
TOILET	
Use a jar or kit to save a gallon per flush. Or, "If it's brown, flush it down. If it's yellow, let it mellow.	Normal flush uses 5-7 gallons
WASHING HANDS	
Full basin: 1 gallon	With tap running: 2 gallons
SHAVING	
Full basin: 1 gallon	With tap running: 20 gallons
BRUSHING TEETH	
Wet brush, rinse: 1 gallon	With tap running: 10 gallons
LEAKS	
Report or fix immediately	Small drip wastes 25 gallons a day

Chapter 29: **MISSION WILD TOOLBOX**

In my experiences working in the developing world, I have found the need to use and also develop appropriate technologies. In a nutshell, this means using affordable materials that are available in a specific area amongst the poor. The following are some examples of appropriate technologies that would be helpful to someone living in a developing country or poor area. I will break it down into the following simple technologies which would enable anyone to take care of their needs for food, clean water, basic health and sanitation:

DRY-LAND FARMING

CONTOUR RIDGES

We have seen a loss of soil in Kenya by people plowing downhill. Much topsoil is lost through the overuse or eradication of basal covering, causing run-off of rain water and soil erosion. In Kenya we had great success in turning a bare arid piece of land into a garden that produced an abundance of vegetables and fruit simply by building simple contours or bunds that would retain water, build topsoil and minimize erosion.

In order to build a bund, the land must be surveyed to mark lines that would follow the contours of the slope. This can be done by constructing a simple A-frame. The steeper the slope, the closer the contour lines should be. Each bund in such an area should be made by digging a trench about a foot in depth and piling the soil up on the lower side of the bund to create a bank. These built up contours help to reduce run-off and erosion. We left the land fallow for one year, and we were amazed to see the development of grasses that covered the area during that period.

After one year, we ploughed in the grasses and planted fruit trees such as papaya along the raised bund. After a few months, we created raised beds four feet wide and fifty feet long. After the first year's planting of vegetables, organic materials such as compost were added to improve the soil structure.

MAKING AN A-FRAME

This simple, easily made and inexpensive surveying tool makes locating the contour lines of your garden area quite doable.

1. The materials needed are:

■ 3 wood or bamboo poles, with 1.5 inch diameter. Two of the poles need to be 2.1 meters long, and the other one 1.2 meters long.

■ Sturdy string for tying or 3 nails

■ A meter of string for hanging weight

■ A rock about the size of your fist or any other similar object of the same weight

2. Make notches 4 inches from one end of the longer poles so that the pieces can fit securely together. Tie or nail the two longer poles securely where they come together so that they will not slip.

3. Spread the legs and brace with the shorter pole to form a figure "A." Tie or nail the crossbar to the longer poles at their mid-point, with about 4 inches of the crossbar extending out past the poles. The crossbar should be parallel to the ground when the A-frame is held up. It supports the legs and will be the guide for marking the level ground position.

4. Firmly tie one end of string to the place where the two tops of the A-frame legs come together.

5. Tie the other end to the rock or weight. The rock or weight needs to be heavy enough that it will not move with the wind. It should hang at least 8 inches below the crossbar.

TO CALIBRATE THE A-FRAME

Place the A-frame on a reasonably level piece of ground in an upright position. Mark the spots where both legs touch the ground. Then mark the crossbar where the string passes it.

Now reverse the position of the A-frame so that you place the legs on the ground exactly opposite from where they were before. Again mark the crossbar where the string falls. If the two marks coincide, then you have found the midpoint on the crossbar and the A-frame is on level ground. If the marks are separate, then mark the point directly between them and retest.

When you have found the midpoint, mark it with a sharp tool.

HOW TO MARK CONTOUR LINES

1. First remove any tall grasses or obstructions to make your work easier. It goes much faster if two people work together. One operates the A-frame while another marks the contour lines.

2. Start at the highest point, at the boundary. Drive the first stake into the ground at the boundary of the area and then place the left leg of the A-frame beside and just above the stake.

3. Place the right leg of leg so that the weighted string passes through the midpoint of the crossbar. Mark this point (and thus the contour) by driving a stake just below the right leg of the A-frame.

4. Now move the A-frame to the right, placing the left leg on the spot where the right leg was before. Position the right leg so the string passes through the midpoint mark. Mark this spot the way you did previously. Continue in this way till you reach the other side of the field.

5. Repeat these steps till you reach the bottom of the hill. The vertical distance between contour lines should be 1.5 meters.

6. When you finish, you will see that some of the stakes will not fit in with the general curve of the contour line because of ground surface irregularities. Just move them to fit in with the general contour.

BUCKET-DRIP IRRIGATION

Being a semi-arid area, we have found the use of bucket-drip kits to be most appropriate and affordable. This is especially so if it is the dry season or if water is scarce or its source is far away.

A drip kit works in this way:

Prepare a bed 4 feet wide by 50 feet long. Mix in compost to provide organic material and improve the soil structure. Make sure that your bed is flat, level and straight. If the beds lie on a small and gradual slope make sure that your water source bucket is on the highest end.

Plant two stout poles with a cross beam at the head of the bed. You can either have the cross beam built to support one bucket that contains at least 5 gallons of water or have your crossbeam constructed to hold two buckets to water two beds at the same time. The bottom of the bucket should be at least 3.3 feet above the surface of the bed. Cut a hole $1^1/_{16}$ inches at the bottom of your

Bucket drip irrigation

plastic bucket. The hole should be in the center, and the appropriate parts are fitted to it to connect two drip lines that are 50 feet long. Each drip line comes with small holes every foot apart. These holes will be where water drips out. Make sure that your drip lines are laid with the holes facing up. Make sure that the bottom end of the drip line is folded over and bound with a thin piece of wire.

The buckets should be filled once in the early morning and once in the late afternoon. Hence not much energy is needed by the user to carry the water only twice a day. Plant your seeds or seedlings at each hole. Covering your bed with mulch will reduce evaporation.

Drip kits are sold at Chapin Living Waters Foundation. (Phone 315 788 0891; Fax 315 782 1490; E-Mail" rchapinw@imcnet.net The kits include comprehensive instructions and diagrams. They do not include buckets which can be readily obtained almost anywhere.

> It is essential that we find a way to make humans more attentive to the earth as well as to all of the earth's residents, present and future. This is the aim of sustainable development, and the responsibility of every individual.
>
> — Christian Broodhag, *Earth from Above*

The drip kits are also sold in Nairobi Kenya by Kenyan Agricultural Research Institute.

A MULTI-STORY GARDEN

In the developing world, there are many people who do not have space to have a garden plot for vegetables. We have found that a multi-story garden, which takes up very little space, can provide a family adequately with vegetables. Watering it is a simple once a day requirement.

How to construct a multi-storey garden:

1. Cut 6 saplings about 4 feet in length and 1 inch diameter. Sharpen a point of the one end.

2. In your selected area, insert a peg in the ground, and with a

Multistory garden

piece of string 18 inches long and a marking stick tied to the end, make a circle on the ground that would have a diameter of 3 feet.

3. At equal spaces, plant your six saplings 9 inches into the ground, using a hammer to establish firmness.

4. In the middle of your circle build up a core of rocks 3 feet tall and with a diameter of about 9 inches. The rocks should be larger at the base than in the pinnacle.

5. Tie string or thin wire at three levels around the saplings.

6. Attach burlap sacking or heavy duty material from meal sacks on the outside of the saplings. You need to sew the sacking on to the sticks at different points. The sacking will then form a strong outside covering.

7. Now cut slits all around the covering at different intervals.

8. Make a mixture of topsoil and compost and very carefully fill the container you have just made. Be careful not to knock over any of your rock tower.

A multistory garden with outside covering attached

Adding compost to a multistory garden

9. Select a gallon-sized can and with nails make holes in the bottom to act as a watering can. You do not need the top. Place the can on the top of the pinnacle of rocks that should just be protruding out of the surface of your soil.

10. Plant seedlings of kale (or preferred vegetable) into the slits on the sides and some on the top surface. You can intersperse various vegetable seedlings in the top or just continue with kale.

11. With a watering can or another container, pour water into the can. The water will now flow down the rock tower and be dispersed to the sides to reach the seedlings as well as the surface area. Your end product will have vegetables growing out the sides as well as on the top of your multistory garden.

MAKING COMPOST

Composting is the natural process of decomposition and recycling of organic material into a humus-rich soil known as compost. Compost provides a low, even release of plant nutrients and stimulates root growth. Composting is a way to recycle waste plant and animal matter as well as kitchen scraps. Do not use meat or bones or dog or cat waste.

1. Compost bins

In countries where old wooden pallets were available, we have made double compost bins with them. To start, obtain seven pallets of the same size. Join the pallets together by driving a metal fencing post or wooden pointed saplings at all the corners into the ground. Slide the pallets onto the stakes and tie the corners top and bottom with wire. You will end up with two bins adjoining each other sharing a common middle section. Below is a photo of a double bin made by our EARS school in Costa Rica.

Double compost bin

2. Pah's Wonder-Working Compost recipe

After the garden near the lake was completely leached out by flooding during El Nino, Christopher Opiyo, affectionately known as Pah, restored it to amazing productivity with his compost made from this recipe.

1. The first layer is of straw or brush about 12 inches thick. This provides a base for the pile.

Compost pit

2. Then the heap is built in layers — first a 2-inch layer of green matter such as weeds, crop wastes, kitchen waste; then a layer of animal manure (less if poultry manure is used) followed by a thin layer of soil.

3. Each layer is watered so as to resemble a squeezed out sponge.

4. The pile is turned after six weeks, and again after 12 weeks to allow air to penetrate all parts of the pile.

The compost should be ready to use in three months. Greener material speeds up composting. Dryer material slows it down.

To check whether or not your compost is cooking, slide a stout stick at an angle into to the compost and leave it there. To test, pull out the stick and feel the end and it should be warm to the touch. When the temperature reaches 100 degrees Fahrenheit, it is ready. Now you can turn it over and start using the cooked compost from the top for your garden.

You need to keep your compost watered, about two or three times a week in dry weather. Do not let it get too wet. If it rains, cover the top with a plastic sheet, a sheet of iron roofing or whatever is available.

3. Pit Method

In countries where no pallets are available, I have used a pit method. Dig two pits, 6 feet by 4 feet and with a depth of 2 feet. Divide the two pits by leaving an earthen section a foot wide in between. You can also add a small wall of either concrete blocks or stone on all the edges. Use the same recipe and procedures as you would for a compost bin.

4. Compost piles

This is probably the simplest, but the most untidy. Using the same recipe, you can just layer your materials on the ground and keep on adding. You will have to turn the pile about once a month. The more you turn it, the sooner it will be ready as turning provides aeration for the microbes at work.

SOLAR COOKERS

Worldwide, 2 billion households depend on wood and charcoal to prepare food. Population growth has gradually depleted the availability of firewood, resulting in swaths of deforested areas.

Nearly one in three people (1.7 billion) live in countries considered to have critically low levels of forest cover, according to a study by Population Action International. Throughout Africa the annual wood consumption for cooking is 1000 pounds of wood per person.

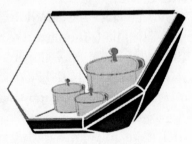

Three million acres of forest in Kenya have been cut down for wood for fuel and charcoal production. Charcoal production, an inefficient conversion of wood to charcoal, has outstripped the remaining forest's ability to regenerate. All this results in soil erosion, depletion of plant and animal life, and forced migration of people who can no longer support themselves on their lands.

Respiratory infection from smoke pollution has become one of the country's leading causes of illness. (Over 5 million children in the third world die from infectious lung disease caused by smoke from cooking fires each year.) Women and children spend a large percentage of their days looking for fuel and cooking. The result is wasted time that could be used in education and other more productive activities and poorer nutrition. Cooking requires an average of 3 kgs. of wood per day per person in rural Africa.

Solar cooking can utilize a safe and free fuel source and make cooking easy and far less time-consuming besides putting less stress on the environment. The production and sale can provide widows with work that sustains them while they have the satisfaction of doing something that helps the community, environment, making life easier for their fellow women.

Our favorite design for solar cooking is that of the Sunstove designed by Richard Wareham from Milwaukee, USA in South Africa through the University of Pretoria. It is inexpensive to make one that will accommodate an average family and so does not need to be subsidized. It is also very tough and lasts well. It is easy to use as it can be faced toward the sun and left for two or three hours before moving its orientation toward the sun again. It cannot overcook and to cook staple foods such as ugali requires 10 to 20 times less woman hours than even on a gas stove. Instead of stirring for 30 minutes, it only takes _ a minute. It requires little or no water for cooking. Also milk and water can be pasteurized by all solar ovens including this one.

Here the students thought we had used magic or tricked them by cooking the cake they taste tested elsewhere than in the solar oven.

GENERAL COOKING INSTRUCTIONS

■ Use ONLY BLACK POTS because they absorb heat. They must not touch the lid of the cooker. The outside of the pot and lid must be black. The inside color of the pot is unimportant.

■ You will need a longer cooking time. For a small amount (1-2 liters) of food, increase time by 50 to 100%. For a large amount (3-6 liters), use two to four times the usual cooking time - especially in less sunny or windy conditions.

■ Food should be in the oven by 10 a.m. so that it catches the best light.

■ Cut in smaller pieces for faster cooking.

■ With the larger amounts of food, a similar cooking period may be used as with an electric slow cooker. Low temperature cooking leaves food more nutritious.

■ Use less water and reduce liquid used in cakes.

■ Cook fresh vegetables without water, or with very little water.

■ Soak dried beans, overnight before use.

■ Food will cook more quickly in two smaller pots than one big one. Minimize the air space between the lid and your food, as this space impedes heat transfer to the food. (The lid of the pot is at the highest temperature.)

CONSTRUCTION OF A SUNSTOVE

There are many kinds of solar ovens which are easier to make. Look up the Cookit on the internet for one you can make in less than an hour. It is basically a reflector used with an oven bag and cooks well though it is clumsier to use than the Sunstove These have been successfully used in refugee camps. Even an old tire with a pane of glass over the top can cook food in a pinch. We enjoy the Sunstove because we find it bakes cakes and bread and cooks meat so well. We even prefer it to a traditional oven.

In Kenya we made a Sunstove with plywood, Plexiglas, and plain metal sheeting. We used old blankets for insulation. We used inner tube for the hinges. Be creative and use materials appropriate to your area.

These plans were taken from the Sunstove internet site. We have used them successfully to make an oven that wowed the local women with its tender meat stews and cakes and bread.

Tools Needed

■ Hand riveter and rivets; industrial stapler and staples
■ Hand staple gun and staples; hammer and large tacks
■ Screwdriver, wood screws, nails, and paint brush
■ Tape measure, wood saw, tin snips, scissors, knife

INSULATION

Bottom: 40mm high-density mineral wool, jute, or fiberglass 50 kg/cu.M or isocyranurate foam. Insulation supports weight of food. Rug or carpet second choice. Scrap textiles, raw wool or cotton require spacers between interior wall and case.

Sides: 40mm low-density fiberglass, mineral wool, jute blanket 12 kg/cu.M will conform to the irregular shape of the sides. Textile cuttings, blanket, rug or carpet will remain vertical on sides.

COVER

TYPE 1 (recommended): Medium-impact acrylic* or polycarbonate sheet, 730x555x1.5mm

With Stops

Stops: Metal or plastic L's, 50mm x 15mm x 1mm thick, two at bottom, one on one side

With Hinge

Weather-resistant, flexible, reinforced, non-fraying material riveted to cover and screwed to frame.

37 15

50

Cover fits under lip

L-type stops

Cover with Top Lip

730mm x 600mm plastic sheet. The extra 45mm forms a lip. Two stops on bottom.

TYPE 2: Thin UV-resistant plastic film or 3mm glass require a frame to hold the film or to protect the glass from breaking. The frame can be held with stops or hinged to the wood frame of the cooker.

TYPE 3: Clear cover, 730 x 555 x 1.5 or 3mm, modified acrylic, polycarbonate, film or glass (must be framed)

Dried lumber 90mm x 20mm cut to length. Paint to protect from weather and warping. Use long screws to hold boards solidly together.

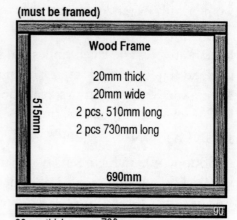

(must be framed)

Wood Frame

20mm thick
20mm wide
2 pcs. 510mm long
2 pcs 730mm long

690mm

20mm thick   730mm

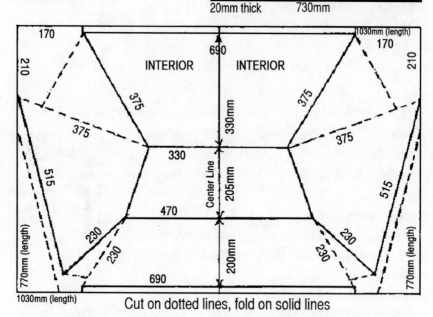

Cut on dotted lines, fold on solid lines

Interior and Case

Note: 30- or 32-gauge galvanized steel, stainless steel, or sheet metal painted with a reflective paint will work. Sheet metal costs and weighs more and is harder to cut and bend than printing plates. Sheet metal does make a stronger case.

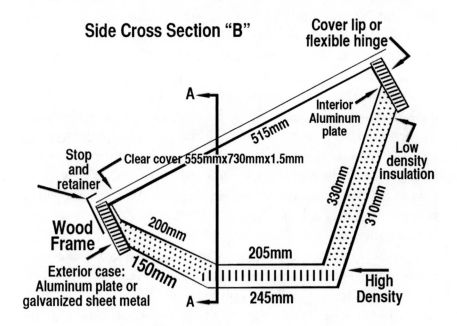

Side Cross Section "B"

Cover lip or
flexible hinge

A

Interior
Aluminum
plate

515mm

Low
density
insulation

Stop
and
retainer

Clear cover 555mmx730mmx1.5mm

330mm

310mm

Wood
Frame

200mm

Exterior case:
Aluminum plate or
galvanized sheet metal

150mm

205mm

245mm

High
Density

A

HAY BOX OR COOKING BASKET

Solar cookers stop cooking when the sun goes down. Families
may eat well after dark. While a well-insulated solar cooker may
retain heat, a haybox or cooking basket
can keep food nicely hot for several hours,
even continuing the cooking process. It can
also be used as a slow cooker that functions
much as a solar cooker.

In Kenya, looking around for what was
locally available to make a cooking basket, we found that there were many larger
inexpensive baskets used in the market
for transporting foods. We found the size
in which a large local cooking pot would fit with a few inches to
spare. We used the local sacks used for rice, maize and beans, which

are a plastic weave to fit in the basket to insulate it and block air flow. Then with around two meters of strong local cloth, we sewed and stuffed three pillows about an inch and a half thick (any available stuffing can be used from clean old rags to dry grass and leaves, but we do not recommend newspaper which sogs up when wet or any material that would release fumes when exposed to heat). We made one round cushion to fit the bottom, a long rectangular one the height of the basket wide and the circumference of the basket long to fit around the inside of the basket, and the third a round pillow to cover the top. It is good to make slips to cover these pillows so that they can be removed for washing. We laid a hot pad of whatever was available on the top of the bottom pillow, on which hot cooking pots could be placed.

To cook using the basket, a pot with rice or beans or whatever needs to be cooked is brought to a rolling boil over heat. Then it is tightly covered with a lid and placed in the basket and quickly covered with the pillow on top. As the basket is well insulated, the heat is retained for several hours, though it slowly diminishes, and the cooking process continues. Rice cooks beautifully in the basket within 30-45 minutes, beans in several hours. Water brought to a boil holds the heat for several hours so that you can make piping hot tea even four hours later.

The hay box works in the same way. Only a hay box is constructed from plywood, and hay is the pillow stuffing. We women in Kenya did not know how to make boxes, but the baskets we could easily do.

These baskets are such a help when one needs to conserve fuel, or one has only one cooking stove with one or two eyes and several different pots of food to cook, or when dried foods are cooked, as no one needs to be around to oversee the long cooking process. In rural Kenya, children are often left to watch the fires and cooking when mothers must go off to work. Often children are burned with these cooking fires. With the cooking basket, a woman can start food, go off to the market or work or church and return to a ready dinner. The advantage of the cooking basket over the solar oven is that it can be locked securely in the house, so that other hungry people are not able to steal the food.

SIMPLE WATER TANK

On our base in Kenya, we have found a very simple method of making water tanks that takes advantage of local skills. We started by cutting long flexible saplings from the surrounding trees and weaving them into a basket like the local Luo granary storage baskets.

Next we built a solid circular foundation from a mixture of stone and concrete on which to set the basket and then to plaster the bottom of the hollow basket into the foundation.

Once the basket is rooted to the foundation, a person enters from the top and applies a layer of fine chicken wire to the basket. This is tied to the basket with thin wire.

From the inside, apply a concrete layer onto the chicken wire. Once this is set, you then apply another coating of cement to the

Simple water tank

outside of the basket. Towards the bottom, fit in a pipe which is threaded on one end to a faucet or tap. Once everything is set, plaster the inside once more with water proof cement.

Lastly, mold a concrete lid that will fit snugly onto the top and once it is set, fit it onto the top of your tank.

For filling, water is either pumped into the top or poured in with a bucket. Rainwater can also be collected from a stand pipe connect to guttering on a metal roof.

TIPPY-TAPS

An ingenious, easily made, simple invention which we got from *Footsteps*, an excellent Christian publication for development from Tearfund, is the tippy-tap. Where there is no running water and water is scarce or hard to fetch, the tippy-tap provides a convenient way to wash hands with a minimal amount of water. It is great to use on camping trips or ministry outreaches in rough areas.

Materials needed are a 2-5 liter plastic container, candle and matches, large nail, cloth and waterproof (nylon or plastic) string. In Kenya we use fishing line or thin, plastic fishing net cord.

Any plastic container of at least 2 liters or quarts will do. We use old vegetable oil containers. First it should be well cleaned. Then, using a nail which has been heated over a candle, poke three holes in a triangle shape close together on the surface below the cap. (Where Ken is pointing in picture above.) Another hole is poked in the handle to allow air flow in as the water comes out of the holes below. If your container does not have a handle, poke two holes to thread the string through so that the tippy-tap will hang by the string in a way that keeps its head up enough for the water in the container to not drip out until it is tilted. You may tie a string around the lid to pull the jug into a tilt and also attach soap to a string tied to those used to hang the tippy-tap. A plastic lid, with a hole in its center, positioned on the string over the soap keeps it from getting wet during rain. We hang the tippy-tap on a tree near the latrines or near the dining hall, so that when people wash their hands, the trees get watered. But in classrooms, we hang the tippy-tap over a basin, which we empty over the garden.

To use the tippy-tap, pull the string to tilt it. Water dribbles out through the holes to wet your hands. Then soap up and tilt again to rinse. We found that more than 30 people can wash their hands on less than 2 liters of water in this way.

Occasionally we put a little bleach in the tippy-tap to kill any algae that may be forming.

COMPOSTING TOILET

The uncomplicated bucket composing toilet is a simple way to handle waste on site. It can be made in a couple of hours. The EARS school near Trinidad made one under the direction of our appropriate tech instructor, Jack Dody, who designed it. Jack tested it by having one in his home. We placed our bucket toilet in the barn which was over a mile away from the nearest restroom, and it was a much appreciated addition.

We made another in Costa Rica at the YWAM Heredia base, which was also used in a location near the gardens and a good walk away from the restrooms.

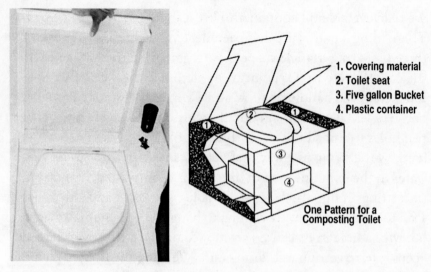

1. Covering material
2. Toilet seat
3. Five gallon Bucket
4. Plastic container

One Pattern for a Composting Toilet

The photo and diagram above are from The Noah Project. Specific instructions are available online at www.Christianhomesteaders.org. The criteria used for its design were:

■ Safety and health, i.e. that it would not contaminate soil or water.

■ Aesthetics — that it would cause no odor nor have unattractive aspects.

■ It would be easily constructed using basic carpentry skills.

■ It would be inexpensive.

■ It would be easily maintained.

■ The system would conserve water.

■ Excrement would be rendered odorless and safe for use as fertilizer.

A composting toilet is a system that converts human waste into organic compost and useable soil. Organic material is broken down as aerobic microbes oxidize carbon in the presence of moisture and air into carbon dioxide gas, converting hydrogen atoms to water vapor.

With the proper conditions, all pathogens are gone and the waste may be used as compost after one year. As a precaution, we use this compost only for trees.

To make the sawdust toilet, first build a frame using eight pieces of 2x4. The length of the frame should be 26 inches, the width 20 inches and the depth 18 inches.

Cut three pieces of thick plywood onto 3 sides, leaving the front open. Cut a fourth piece of plywood to cover the top. Before nailing it on, mark two holes by tracing round the bottom of the bucket so that you have a hole as the entrance over the depository bucket. Mark a second hole using the same method for your sawdust container. The holes should be alongside each other about 6 inches apart.

Using a hole saw, cut out the holes at a small slant (so that the piece can become the lid and not fall through!) and file the edges so that they are not rough. Nail this top cover onto the top of the box. Use the two pieces that you cut out as covering of the holes when not in use.

Screw a knob onto the center of each hole cover for easy lifting and replacing. Paint the wood with exterior acrylic paint for easy washing. Slide the two buckets into position under the holes. One of them is your toilet, and the other is the container for either sawdust or dry leaves.

Before using the toilet, sprinkle a layer of sawdust or leaves on the bottom of the bucket. Cover urine or feces with sawdust by using a small container to dip into the sawdust. It takes plenty of covering material to assure that the composting material is not too wet and to completely cover the contents of the toilet . The toilet does not give off an odor or attract flies.

Use regular white toilet paper which provides cellulose for the soil.

When the toilet bucket is close to full, slide it out via the front and take it outdoors to a prepared area for composting it separate from your other compost bin or pile.

A simple fenced area 4 feet high and 3 feet across or an area made with concrete block walls of similar dimensions with 3 sides will do. Cover with a layer of grass clippings, leaves or dirt.

SOLAR WATER DISINFECTION: SODIS METHOD

The SODIS is simple technology for purifying small amounts of water to make it of good enough quality to drink by using solar radiation and heat to destroy pathogenic microorganisms in it. If water has been contaminated by chemicals, it will not remedy that condition. Water that is cloudy must be filtered or allowed to settle so that you could read large letters through a few inches of it before it can be disinfected using the SODIS method. This method will not work for large amounts of water.

Fill a 2-liter clear plastic bottle (Not bluish. Clear PET plastic is preferred.) with water to 85% of its capacity. Shake the bottle for 20 seconds to aerate it. Place it in the full sun for six hours on a sunny or partly cloudy day. For totally cloudy days, it must be put out two consecutive days. If it is rainy, collect rainwater for drinking!

If the water is heated up to 122 degrees Fahrenheit or 50 degrees Celsius, the purification rate speeds up 300%. Placing the bottle on a metal roof or black surface or painting the bottom of the bottle black, will help to raise the temperature in it.

We tried this using water from the very polluted Lake Victoria during a cholera epidemic. When tested, the water was found to be pure and good for drinking.

COLLOIDAL SILVER

Colloidal silver is water containing both microscopic particles of elemental silver and silver ions in suspension. It is a powerful, natural antibiotic which disables the enzyme that one-celled bacteria, viruses and fungi need for oxygen metabolism. Disease-causing organisms are destroyed without harm being done to the human body.

Having colloidal silver on hand has been a wonderful help to us in a place where germs seem vicious and thriving. We have seen quick healing result when it is applied topically to infections on the skin, and it has always seemed to work internally as a cure for diarrhea. It is important to exercise care in making colloidal silver

so that it will be pure and
the particle size small.
The smaller the particle
size, the more likely the
colloidal silver will be absorbed in a useable form
by the body.

Materials Needed:

■ Four 9-volt alkaline
batteries.

■ Two 9-volt battery-to-
clamp adapters, and two
metal clamps.

■ Glass container that
holds 8 ounces of water

Inactivation of microorganisms by UV-A-radiation and thermal treatment

■ Two fine silver (not sterling silver) rods, minimum 14 gauge or
two strips of 18 gauge fine silver wire. It needs to be at least 99.9%
pure.

■ Good quality distilled water, at least 10 ounces

■ Pure, white paper towels.

■ One absolutely clean nylon scrub pad

■ A glass container for storing the colloidal silver. It is better to
have UV protected or dark glass.

Directions:

■ Wash hands.

■ Clean the inside of the glass container with the nylon scrub pad,
dry it with a clean paper towel and then rinse it with a little distilled water. Put 8 ounces of distilled water in the glass container.

■ Use the nylon scrub pad to rub the silver wires to make them
smooth and clean. Then wipe them with a clean paper towel barely
damp with the distilled water.

■ Attach one metal clamp to the red (positive) wire of the
nine-volt battery-to-clamp adapter, and one clamp to the black (

negative) wire of the second lead adapter. Make sure the wires are stripped enough for a good connection.

■ Attach two of the 9-volt batteries together, negative to positive. Then do the same with the remaining two batteries.

■ Just before using, attach the two sets of batteries together, with negative to positive again.

■ Finally attach the positive (red) clamp/lead to the exposed positive pole on the batteries and the negative (black)clamp/lead to the exposed negative pole of the batteries.

■ Place the two silver wires or rods parallel between one and a half to two inches apart near the center of the glass container, not allowing them to touch each other or the sides or bottom of the container. (you will have to rig up something to keep them in place.)

■ Attach a clamp to each rod or wire.

■ Watch the wires/rods. After 10 to 15 minutes, a narrow yellow stream should begin to drift up from the wires/rods. After 5 more minutes, remove the negative rod from the water carefully. Then in a couple of minutes carefully remove the second wire.

■ Take the batteries apart.

■ Keep the colloidal silver in a glass container with a lid. It is better to keep it from being exposed to sunlight.

The purity and concentration may vary quite a bit, but this colloidal solution should be both safe and effective. Two ounces of colloidal silver in a glass of water can be drunk for internal micro-organisms or the solution made can be applied topically.

Chapter 30: **CONCLUSION**

"Now hope does not disappoint, because the love of God has been poured out in our hearts by the Holy Spirit who was given to us."
— Romans 5:5

I n this book I have attempted to tell something of my journey through life and how Nancy and I have been fulfilled through the adventure of it. I have also tried to lay out some teachings on the stewardship of God's creation in the hope that others may take over this mantle. I have attempted to show how important this type of ministry is towards fulfilling The Great Commission in discipling the nations.

We are not at the end of our journey yet, and much more needs to be played out before the completion of it. We live in perilous times, but also exciting times, in which we can all be part of the cast in welcoming the King. Mankind at the moment seems to be bent on separating himself from this King, our Creator and God. The Church is fragmented and not a cohesive body of believers that it should be, that is living in unity, but seems so divided in isms. God's intent for His Church, His Bride, is to be one, busy in ushering in the Kingdom of God and not the kingdom of a denomination!

Nancy and I in our older age are still involved in missions. We still travel abroad to come alongside ministries we have helped birth and assist new ministries in Creation Stewardship in different parts of the world. Kenya and especially the Nyanza Province of which the Suba District is a part are still much on our hearts, as is Maasailand. This is a focus we will not give up. There is so much yet to be done. We cannot sit back in complacency and watch thousands of people suffer in extreme poverty, disease, corruption and early demise. Each life is a potential for God.

In July of 2008, we are heading back to the Suba District to help the ministry in that district become more effective, and by doing so we hope to be part of reaching 500,000 people in the Nyanza Omega zone within the next ten years. Launching a revised ministry which we are calling Suba Environmental Education of Kenya (S.E.E.K), we plan to convert most of the base grounds into a nature reserve building permanent camping cabins.), and the education center to a Biblical Ecology Study Center. (The rest will be a demonstration garden.) Here a camping program will be run for the children from Christ's Gift Academy as well as students from the other schools in the area. We will also run courses for teachers, pastors and community members. Part of the year will be open to visitors from other parts of Africa and overseas. Participants will be exposed to the beauty and wonders of God's creation through environmental education, adventure and exploration, so that they may know God and experience His wonderful attributes, while becoming environmentally literate, able to address environmental issues in a Biblical way that cares for people and cares for the earth.

By Biblically based, scientifically accurate, comprehensive, quality environmental education programs, we believe to develop environmentally literate children, youth and adults who have the skills, knowledge, and inclinations to make well-informed choices. We hope they will be enabled to become wise stewards of God's creation, being sensitive to the needs of man, cognizant of man's responsibility to the life-supporting systems, and having Biblical understanding of how man's survival depends upon the conservation of his natural and man-made resources. They will, therefore, be able, as they depend on God and live in faith, to use their God-given resources wisely so that they can enjoy God's provision and serve others.

Our focus will be the love of God, His love for people, His church, and His creation.

Our students, young and old, will use the wilds and rural Africa as a real-world context for learning, linking their classroom learning or professional understanding to natural and man-made Kenyan environment. They will engage in hands-on learning that increases their knowledge and awareness about the environment. The methods used will encourage inquiry and investigation, develop critical thinking skills and problem-solving decisions, and use teamwork. As environmentally literate students they will be able to make wise decisions in an ever changing world, able to understand God's ways and natural principles, thus positively affecting their lives and those of their community.

Here are some recent facts concerning Nyanza Province as I pen this final chapter in March 2008:

■ A recent survey of Kenya lists Nyanza as the poorest province in the country, with poverty levels ranging from 65 to 80 percent.

■ The Fourth Kenya Human Development Report in early 2007 showed that Nyanza had the lowest life expectancy, 46 years.

■ The Kenya Demographic and Health Survey says that only 0.6 percent of Nyanza residents have access to piped water despite the presence of massive water resources in the area, mainly from Lake Victoria.

■ Only 5.1 percent of the people have access to electricity.

■ Nyanza reports high child-mortality rates, 206 deaths per 1000 births.

■ The province has the highest HIV/AIDS prevalence rate in Kenya.

■ The large amount of orphans is compounded by the fact that most of them do not have grandparents.

■ An average 64 percent of the people are living on less than Sh 40 a day. That is 57 U.S. cents a day. The poverty levels are taking toll on education, especially in female-headed households.

■ Food production has decreased alarmingly over the years.

■ The Suba District which gives about 70 percent of fish for export has extremely poor infrastructure. It has not one inch of paved roads, no bank, no district hospital and only one qualified doctor. It also leads in HIV prevalence in the whole country, 42 percent on the mainland and on some of the islands of Lake Victoria as high as 70 percent.

■ Suba District is home to the Ruma National Park. Tsetse fly in the park has killed some 18,000 domestic animals in the surrounding areas in the last few years. This denotes a high financial deficit for the local people. Work needs to be done in eliminating the Tsetse fly.

The list goes on and on. There is a huge lack of clean water, causing numerous water-borne diseases, killing off people in high numbers. Malaria rates are huge. It has been estimated that every Nyanza resident either is suffering from malaria or has had it in the past. Thousands of people are dying annually from malaria. Witchcraft still exists.

Through our work with YWAM Kenya Islands Base, in co-operation with government agencies, visiting mission teams, and doctors, we have possibly slowed down the rate of disease and death. We distributed close to 8,000 treated mosquito nets to families in 2006, with the goal for every family in the Suba District to have a number of nets together with a Bible. Orphans are being looked after, and people are giving their lives to Jesus, churches are working on environmental issues and care of widows, but there is a long way to go. .

I am praying that this book might be a catalyst for more young people to become involved in missions, especially those with science degrees or a science background who seek to use their knowledge and zeal towards rehabilitating the land and redeeming the people, through discipleship and environmental stewardship in countries such as Kenya. I am praying for a whole army of young people going out into the world, fulfilling The Great Commission and heralding in the Kingdom of God on earth.

I have found richness in the Son, the King of Kings, and my hope is that many more in this world will find it too.

Amen.

ENVIRONMENTAL GLOSSARY

Abiotic factor: a non-living component of the environment, such as soil, nutrients, light, fire, or moisture.

Adaptation: the ability, through inherited structural or functional characteristics, that improves the survival rate of animal or plant life in a particular habitat.

Agro ecosystem: an agricultural system understood as an ecosystem.

Agro forestry: the practice of including trees in crop-or animal-production.

Aquaculture: the production of food and feed using aquatic systems.

Autotroph: an organism that satisfies its need for organic food molecules by using the energy of the sun, or of the oxidation of inorganic substances, to convert inorganic molecules into organic molecules. Green plants are autotrophs.

Bacteria: single-celled microscopic organisms found in every habitat, ecosystem or cycle.

Biomass: the mass of all organic matter in a given system at a given point in time.

Biotic factor: an aspect of the environment related to organisms or their interactions.

Browser: an animal that feeds on the leaves and stems of plants other than grass.

Buffer zone: a less-intensively-managed and less-disturbed area at the margins of an agro ecosystem that protects the adjacent natural system from the potential of negative impacts of agricultural activities and management.

Carnivore: an animal that lives by eating the flesh of other animals.

Carrion: dead and decaying flesh of animals.

Climax: in classical ecological theory, the end point of the successional process.

Commensalism: an inter-organism interaction, in which one organism is aided by the interaction and the other is neither benefited nor harmed.

Community: all the organisms living together in a particular location.

Competition: an interaction in which two organisms remove from the environment a limited resource that both require, and both organisms are harmed in the process. Competition can occur between members of the same species and between members of different species.

Compost: mixed decayed and decaying organic matter with available nutrients useful for fertilizer.

Conservation: the wise use of the earth's natural resources that ensures their continuing availability for generations to come.

Consumer: an organism that ingests other organisms or their parts or products to obtain its food energy.

Cross-pollination: the fertilization of a flower by pollen from the flower of another individual of the same species.

Decomposer: a fungal or bacterial organism that obtains its nutrients and food energy by breaking down dead organic and fecal matter and absorbing some of its nutrient content.

Detrivore: an organism that feeds on dead organic and fecal matter.

Diversity: the number of variety of species in a location, community, ecosystem, or agro ecosystem.

Dominant species: the species with the greatest impact on both the biotic and abiotic components of its community.

Drought: an indefinite period of time when little or no rain falls on an area.

Ecological diversity: the degree of heterogeneity of an ecosystem's or agro ecosystem's species makeup.

Ecological niche: an organism's place and function in the environment, defined by its utilization of resources.

Ecology: the study of the relationships of living things to each other and to their non-living environment.

Environment: the term which describes all external conditions such as soil, water, air and organisms, surrounding a living thing.

Epiphyte: a plant that uses the trunk or stem of another plant for support, but that draws no nutrients from the host plant.

Erosion: the weathering of the earth's surface by water, wind, ice and other natural forces.

Evapotranspiration: all forms of evaporation of liquid water from the earth's surface, including the evaporation of bodies of water and soil moisture and the evaporation from leaf surfaces that occurs as part of transpiration.

Grass: a plant characterized by jointed, sometimes hollow stems, two-part leaves consisting of a sheath around the stem and a long, flat blade, tiny flowers on a small spike and dry, seed like fruits.

Grazer: an animal that feeds on grass such as zebra.

Habitat: the particular environment, characterized by a specific set of environmental conditions, in which a given species occurs.

Herbivore: an animal that feeds exclusively or mainly on plants. Herbivores convert plant biomass into animal biomass.

Heterotroph: an organism that consumes other organisms to meet its energy needs.

Horizons: visually distinguishable layers in the soil profile.

Host: an organism that provides food or shelter for another organism.

Humus: the fraction of organic matter in the soil resulting from decomposition and mineralization of organic material.

Hydrological cycle: the process encompassing the evaporation of water from the earth's surface, its condensation in the atmosphere, and its return to the surface through precipitation.

Insolation: the conversion, at the earth's surface, of short wave solar energy into long wave heat energy.

Inversion: the sandwiching of a layer of warm air between two layers of cold air in a valley.

Legume: a plant in the Leguminosae family. Most of these species bear a bean or pea type pod, and can fix nitrogen.

Mammals: the term for the group of animals including human beings that are all warm-blooded, have milk-producing glands, are partially covered with hair and normally bear their young alive.

Microclimate: the environmental conditions in the immediate vicinity of an organism.

Mutualism: a relationship between two organisms either plant or animal in which both partners benefit.

Organic matter: materials containing molecules based on Carbon, usually referring to soil organic matter.

Organism: an individual of a species.

Overgrazing: intensive feeding on the vegetation on an area by wild or domestic animals which causes serious and often permanent damage to the area's plant life.

Oxidation: The loss of electrons from an atom that accompanies the change from a reduced to an oxidized state.

Parasite: an organism that uses another organism for food and thus harms the other organism.

Parasitism: an interaction in which one organism feeds on another organism, harming but generally not killing it.

Population: a group of individuals of the same species that live in the same geographic area.

Predation: an interaction in which one organism kills and consumes another.

Predator: an animal that consumes other animals to satisfy its nutritive requirements.

Prey: a living animal that is captured by a predator such as a lion, leopard, or cheetah for food.

Pride: a family or group of lions.

Primary production: the amount of light energy converted into plant biomass in a system.

Primary succession: ecological succession on a site that was not previously occupied by living organisms.

Renewable resources: resources that are capable of being regenerated or replaced by ecological processes on a time scale relative to their use. These resources such as biomass or energy from animal traction, are contrasted with non-renewable resources, such as fossil fuels and mined products.

Scavenger: an animal like the vulture or hyena that lives by devouring the dead remains of other animals.

Secondary succession: succession on a site that was previously occupied by living organisms but that has undergone severe disturbance.

Shrub: a mall woody plant with more than one stem rising from the ground.

Slash and burn: a type of shifting cultivation that uses fire to clear fallow areas for cropping.

Species: the term is singular or plural, and relates to a group of plants or animals with common characteristics.

Succession: the process by which one community gives way to another.

Symbiosis: a relationship between different organisms that live in direct contact.

Territory: an animal's domain which it defends against species of its own kind or other species, or an area used by a particular animal for feeding and breeding.

Transpiration: the evaporation of water through the stomata of a plant, which causes a flow of water from the soil through the plant and into the atmosphere.

Trophic level: a location in the hierarchy of feeding relationships within an ecosystem.

Veld: a term used in Africa for open land used for grazing and other needs.

Xerophyte: a plant that is adapted to growing under very dry conditions and extended droughts.

BOOKS ON ENVIRONMENTAL STEWARDSHIP

Aeschliman, Gordan and Campolo, Tony. 1992. *50 Ways You Can Help Save the Planet*. Downers Grove, Illinois, Intervarsity Press.

Bhagat, Shantilal P. 1990. *Creation in Crisis: Responding to God's Covenant*. Elgin, IL: Brethren Press.

Bratton, Susan Power. 1992. *Six Billion and More: Human Population Regulation and Christian Ethics*. Louisville, KY: Westminster/John Knox Press.

DeWitt, Calvin B., 1991. *The Environment and the Christian: What can We Learn from the New Testament?* Grand Rapids, MI: Baker Book House. With contributions from Calvin B. DeWitt, Raymond C. van Leeuwen, Ronald Manahan, Vernon Visik, Loren Wilkinson and David Wise.

Evans J. David, Ronald J. Vos, Keith P. Wright, Editors 2003. Pasadena CA. William Carey Library.

Freudenberger, C. Dean. 1990. *Global Dust Bowl: Can We Stop the Destruction of the Land Before It's Too Late?* Minneapolis, MN: Augsburg Fortress.

Gelderloos, Orin.1992. *Eco-Theology: The Judea-Christian Tradition and the Politics of Ecological Decision Making*. With a foreword by Kathy Galloway. Glasgow, Scotland: Wild Goose Publications.

Granberg-Michaelson, Wesley, ed. 1987. *Tending the Garden: Essays on the Gospel and the Earth*. Grand Rapids, MI: Eeerdmans.

Granberg-Michaelson, Wesley. *Ecology and Life: Accepting Our Environmental Responsibility*. 1998. Waco, TX: Word Books.

Hall, Douglas John. 1987. *Imaging God: Dominion as Stewardship*. Grand Rapids, MI: Eeerdmans.

Kennedy, Dr. D. James and Beisner, Dr. E. Calvin. 2007. *Overheated: A Reasoned Look at the Global Warming Debate*. Fort Lauderdale, FL: Coral Ridge Ministries Media Inc.

Miller, Darrow L. 1998. *Discipling Nations: The Power of Truth to*

Transform Cultures. Seattle, WA: YWAM Publishing.

Nash, James A. 1991. *Loving Nature: Ecological Integrity and Christian Responsibility*. Nashville, TN: Abingdon Press.

Prance, Ghillean T., and Calvin B. DeWitt, eds, 1992. *Missionary Earthkeeping*. Macon, GA: Mercer University Press.

Robinson, Tri, With Jason Chatraw. 2006. *Saving God's Green Earth: Rediscovering the Church's Responsibility to Environmental Stewardship*. Norcross, GA: Ampelon Publishing.

Saint, Steve. 2001. *The Great Ommission*. Seattle, WA: YWAM Publishing.

Wilkinson, Loren, ed. 1991. *Earthkeeping in the Nineties: Stewardship and the Renewal of Creation*. Grand Rapids, MI: Eerdmans.

Wilkinson, Loren and Mary. 1992. *Caring for Creation in Your Own Backyard*. Ann Arbor, MI: Servant Publications.

Young, Richard A. 1994. *Healing the Earth: A Theocentric Perspective on Environmental Problems and Their Solutions*. Nashville, TN. Broadman & Holman.

ORGANIZATIONS INVOLVED IN CREATION CARE

IN THE UNITED STATES

A Rocha USA
P.O. Box 1338
Fredericksburg, TX 78624
Phone: (830) 992-7941
E-mail: usa@arocha.org

American Scientific Affiliations
P.O. Box 668
Ipswich, MA 01938
Phone: (978) 356-5656

Au Sable Institute of Environmental Studies
Administrative Office
3770 Lake Drive SE
Grand Rapids, MI 49546
Phone: (616) 526-9952
Fax: (616) 526-9955
E-mail: administration@ausable.org

Christian Environmental Association (Target Earth)
Target Earth
P.O. Box 10777
Tempe, AZ 85284
Phone: (610) 909-9740
E-mail: infor@targetearth.org

Christian Environmental Studies Center at Montreat College
North Carolina
Phone: (800) 622-6968 or (828) 669-8012
E-mail: admissions@montreat.edu

The Cornwall Alliance for the Stewardship of Creation
9302-C Old Keen Mill Road
Burke, VA 22015
Phone: (703) 569-4653
Web site: www.cornwallalliance.org

Creation Studies Institute
2002 W. Cypress Creek Rd, Suite 220
Ft. Lauderdale, FL 33309
Phone: (800) 882-0278 or (954) 771-1652
Web site: www.creationstudies.org

Earth Ministry
6512 23rd Ave. Ste 317
Seattle, WA 98117
Phone: (206) 632-2426

ECHO (Environmental Concerns for Hunger Organization)
17391 Durrance Road
N. Fort Myers, FL 33917
Phone: (239) 543-3246
E-mail: echo@echonet.org
Web site: www.echonet.org

Evangelical Environmental Network, USA
4485 Tench Road, Suite 850
Suwannee, GA 30024
Phone: (678) 541-0747
E-mail: een@creationcare.org

Floresta
4903 Morena Blvd. Suite 1215
San Diego, CA 92117
Phone: (858) 274-3718 or (800) 633-5319
Web site: www.Floresta.org

The National Religious Partnership for the Environment
49 South Pleasant Street, Suite 301
Amherst, MA 01002
Phone: 413 253 1515
E-mail: nrpe@nrpe.org

Rediscovering the Outdoors
George Fox University
414 N. Meridian Street
Newberg, OR 97132
Phone: (503) 538-8383

Web of Creation
Ecology Resources to Transform Faith and Society
E-mail: webofcreation@lste.edu

IN COSTA RICA

YOUTH WITH A MISSION (YWAM)
Environment and Resource Stewardship School (EARS)
JCUM, Heredia. Apdo 1444-3000
Heredia 3000, Costa Rica
Phone: 011 506 2267 7063
E-mail: terrylkeith@gmail.com
Website: www.ywamconnect.com/sites/ywamheredia

IN KENYA

Care of Creation, Inc.
Attention: Craig Sorley, Director (Kenya)
Brackenhurst
P.O Box 32
Limuru
Kenya
E-mail: ctsorley@att.net
Phone: 011 254 66 73007
Mobile: 254 733 451372
Web site: www.careofcreation.org

SUBA Environmental Education of Kenya
A Ministry of Kenya Islands Missions, Inc.
U.S. address:
810 Leisure Lane
Gatlinburg, TN 37738
Phone: (865) 436-4114
E-mail: ywamwild@aol.com

Mwamba
Colin and Ronnie Jackson
P.O. Box 383
Watamu 8020
Kenya
Phone: 254 0722 842366 or 254 042 32023
E-mail: colin.jackson@arocha.org
Web site: www.arocha.org

Walking With Maasai
P.O. Box 530
Narok
Kenya
Phone: +8821 6510 73 287
Web site: www.walkingwithmaasai.org

MORE U.S. ORGANIZATIONS

APPROPRIATE TECHNOLOGY, SUSTAINABLE AGRICULTURE, HEALTH

Equip International
P.O. Box 1126
Marion, N.C. 28752-1126
Phone: (828) 738-3891
Web site: www.equipministries.org

The Noah Project
c/o Jack Dody
P.O. Box 26
Rush, CO 80833
Phone: (719) 360-3075
E-mail: jack@christianhomesteaders.org
Web site: www.Christianhomesteaders.org

University of the Nations
College of Science and Technology
Seminars and Workshops
Coordinator, Andrew West
University of the Nations - Kona
75-5851 Kuakini Highway #111
Kailua-Kona, HI 96740 USA
Phone: (808) 326-4431
Fax: (808) 326-4443
Email: colsat@uofnkona.edu

Water for Life Institute
75-5851 Kuakini Hwy #75
Kailua Kona, HI 96740 USA
Phone (808) 326-4420
FAX: (808) 326-4443
E-mail: info@waterforlife.org

BUILDING

Habitat for Humanity International
121 Habitat Street
Americus, GA 31709-3498
Phone: 912 924 6935
E-mail: info@habitat.org

ENVIRONMENTAL EDUCATION

Genesis Account American Outdoor Schools
48899 Winners Circle
Coarsegold, CA 93614
Phone: (559) 642-2435
E-mail: bfrem@sierratel.com

FEEDING THE HUNGRY

Gleanings for the Hungry
A Ministry of YOUTH WITH A MISSION
P.O. Box 309
Sultana, CA 93666
Phone: (559) 591-5009
E-mail: gleanings@attitude.com
Web site: www.gleanings.org

Heifer Project International
1 World Avenue
Little Rock, AR 72202
Phone: (800) 422-0474
Web site: www.heifer.org

HEALTH AND PARTICIPATORY LEARNING

Lifewind International (CHE)
P.O. Box 576645
Modesto, CA 95357-6645
Phone: (883) 403-0600 or (209) 534-0600
E-mail: info@LifeWind.org
Web site: www.LifeWind.org
WATER AND DISASTER

Lifewater International
P.O. Box 3131
San Luis Obispo, CA 93403
Phone: (805) 541-6634
Web site: www.lifewater.org

Living Waters for the World
121 County Road 422
Water Valley, MS 38965
Phone: (662) 234-5705

And there are many more organizations doing excellent work internationally.

BIBLIOGRAPHY

Arthus-Bertrand, Yann. 2002. *Earth From Above*. Published by Harry N Abrams, Inc. New York, NY.

Beisner, Dr. Calvin, Dr.E. and Kennedy, Dr. D. James. 2007. *Overheated: A Reasoned Look at the Global Warming Debate*. Coral Ridge Ministries. Fort Lauderdale, Florida.

Cunningham, Loren, 2007. *The Book that Transforms Nations*. YWAM Publishing, Seattle WA 98155.

Eldredge, John. 2001. *Wild at Heart*. Thomas Nelson, Inc. Nashville, TN.

Evans, David. J, Vos, Ronald, J. Wright, Keith, P. Editors. 2003. *Biblical Holism and Agriculture*. Published by William Carey Library. Pasadena, CA.

Lucado, M., *In the Grip of Grace*. Thomas Nelson, Inc. Nashville, TN.

Miller, Darrow L. 1998. *Discipling Nations*. YWAM Publishing. Seattle, WA.

Odum, Eugene Pleasants. 1983. *Basic Ecology*. Saunders College Publishing. Fort Worth, TX.

Richards, D. 1985. *An Alternative Program for Developing Environmental Literacy in 12 Year Old South African School Children*. M.Ed Thesis, University of Natal.

Richards, D and Walker, C. 1986. *Walk Through the Wilderness*. Published by the Endangered Wildlife Trust and the Wilderness Trust of Southern Africa.

Robinson, Tri, 2006. *Saving God's Green Earth*. Ampelon Publishing, Norcross, GA

Saint, Steve. 2001. *The Great Omission*. YWAM Publishing. Seattle, WA.

Sheets, Dutch.2001. *Praying For America*. Regal Books Gospel Light. Ventura, CA.

Smith, Robert L. 1990. *Ecology and Field Biology*. Harper Collins Publishers. New York, NY.

PHOTO CREDITS

Leading a wilderness trail outing in Umfolozi . . . . Andre Brink

The rumbling Loita Hills, 7,000 feet above sea level . . . . Don Richards

A course wilderness trail in the St. Lucia area . . . . Don Richards

A group of Boy Scouts observing animals in Ruma National Park . . . . Hennie Marais

The author viewing white rhinos in the Maasai Mara . . . . Andre Brink

Typical African scavengers: the white-backed vulture . . . . Andre Brink

Observing elephants during an eco-tour of the Maasai Mara . . . . Don Richards

Ugly but beautiful: warthogs drinking . . . . Don Richards

Maasai children dressed traditionally for a wedding . . . . Andre Brink

Maasai village in the Loita Hills . . . . Don Richards

Jarawara tribespeople in the Brazilian Amazon . . . . Don Richards

Maasai warriors in traditional dress . . . . Andre Brink

The author teaching adaptation in the Loita Hills . . . . Kashu Parit

Researching water quality on Lake Victoria . . . . Nancy Richards

The author showing hippo tracks to Zulu youth . . . . Don Richards

CONTACT INFORMATION

TRANSFORMATION THROUGH BIBLICAL ENVIRONMENTAL EDUCATION

Mission Wild: One Man's Adventures in Saving the Planet: www.missionwild.com
Contact the author: ywamwild@aol.com or write to Don and Nancy Richards, 810 Leisure Lane, Gatlinburg, TN 37738
E-mail the publisher: info@missionenablerspublishing.com or call (808) 896-2515 for bulk orders or publishing services.

Mission Enablers International (MEI) is a 501-c-3 nonprofit organization.
Visit www.missionenablersinternational.com for opportunities to be involved in serving or sponsoring this and other ministries.

Kenya Islands Missions, Inc.
109 Doral Lane,
Hendersonville, TN 37075
Phone: (615) 824-7480
E-mail: gljwmorgan@bellsouth.net

Mission Wild
Sherry and Bill Simpson
101 Willis Drive
Hendersonville, TN
(615) 824-9776 or (615) 668-6260

Locations and dates for the Environment and Resource School (EARS) are available online at: www.ywamconnect.com/sites/ywamheredia or www.mission-wild.com
The EARS school is part of a degree program from the University of the Nations, Kailua-Kona, HI. Credits are accepted at more than 2,000 colleges and universities around the world. Information available online at: www.uofnkona.edu
The Discipleship Training School (DTS) is a prerequisite for the EARS program. The EARS program can be taken in several locations internationally. Field assignments are determined by the individual locations program director. The DTS and EARS are accredited from Global Accreditation Association (GAA) through YWAM's University of the Nations, Kona, HI.

Don and Nancy Richards are the parents of five adult children and spend the major part of each year at training locations. Although he is a U.S. citizen, Don is presently director of Suba Environmental Education of Kenya in Africa and executive director of Walking with Maasai in Kenya. He is currently chairman of the Kenya Islands Mission in the United States. He was a game warden in The Umfolozi Game Reserve in Zululand as well as a wilderness trails officer with the Wilderness Leadership School in South Africa for many years, leading people on foot trails in big-game country and sleeping out with them for three nights or more in the bush. He has pioneered environmental education in South Africa as well as starting up the first environmental centers there.

Thank you for your prayers and sponsorship of this and other great MEI ministries.
Your tax-deductible donation can be processed online at:
www.missionenablersinternational.com

God bless you!

LaVergne, TN USA
08 January 2011
211606LV00003B/6/P